Availability Analysis:
A Guide
to Efficient
Energy Use

2-3 The Second Law of Thermodynamics, Entropy

that must be met only by real heat engines. The Kelvin–Planck statement requires that it be satisfied by all heat engines, however idealized they may be. No heat engine, real or ideal, can be 100% efficient. Given that no heat engine can have a thermal efficiency of 100%, it is of interest to determine the maximum *theoretical* efficiency. This is considered below for an important special case in the subsection on the *Carnot efficiency*, but first some additional important concepts relating to the second law are presented.

Reversible and Irreversible Processes. A process is said to be reversible if it is possible for its effects to be eradicated in the sense that there is some way by which both the system and its surroundings can be exactly restored to their respective initial states. A process is irreversible if there is no way to undo it. That is, no way by means of which the system and its surroundings can be exactly restored to their respective initial states. Alternatively, a process is irreversible if the undoing of all its effects makes possible the construction of a PMM2. This test is illustrated in the example to follow.

Example 2-1. Consider the case of a spontaneous heat transfer of energy through a rod whose ends are held at constant temperatures T_1 and T_2 ($T_2 < T_1$), while the rod is insulated along its length and its state does not change with time. Show that spontaneous heat transfer through the finite temperature difference is an irreversible process.

Solution. The actual, spontaneous process is shown in Fig. E2-1a. If the process were reversible, energy in amount q could pass unaided from the body at T_2 to the body at $T_1 > T_2$. However, by introduction of a heat engine, a system could be obtained which operates in a cycle while taking up energy in net amount $(q - q_R)$ from a single source (the body at T_2) and converting it entirely to work. The system, shown by a dashed line in Fig. E2-1b, is a PMM2.

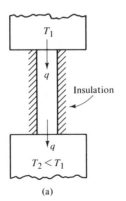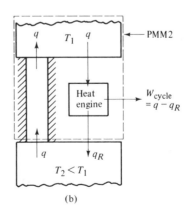

Figure E2-1

This possibility is denied by the Kelvin–Planck statement of the second law. It can be concluded, then, that the spontaneous process is irreversible.

In addition to spontaneous heat transfer through a finite temperature difference, there are many other effects that render processes irreversible. Familiar examples are *friction*, including sliding friction in mechanical devices and friction associated with the flow of viscous fluids, *unrestrained expansion* of fluids, *spontaneous chemical reaction*, *mixing of two different substances*, and *electric current flow through a nonzero resistance*. Effects such as these are termed for easy reference "irreversibilities."

Internal and External Irreversibilities. Irreversibilities can often be divided into two classes, internal and external. Internal irreversibilities are those which occur within the control mass or control volume, while external irreversibilities are those which occur within the surroundings, normally the immediate surroundings. As this division depends on the location of the control surface there is some arbitrariness in the classification (by locating the control surface to take in the immediate surroundings, all irreversibilities are internal). Nonetheless, as shown by subsequent developments valuable insights frequently result when this distinction between irreversibilities is made.

When irreversibilities are absent from within a control mass during a process, the process is said to be *internally reversible*. In such a process, the control mass passes through a sequence of equilibrium states referred to as the *path*.

As an example of the use of the internally reversible process idea, an expression for the work done as a fluid experiences an internally reversible expansion or compression in a piston–cylinder assembly will now be developed. The work is given by

$$W = \int_1^2 F \cdot dx$$

where F is the force of the fluid on the piston. Since there is internal balance at each step of the process, the force at a given instant is $p\alpha$ where p is the uniform fluid pressure at that instant and α is the area of the piston face. Recognizing αdx as the change in volume

$$W = \int_1^2 p\,dV \qquad (2\text{-}3a)$$

As shown in Fig. 2-2, when the states visited in the process are represented by a curve (a path) on a plot of pressure versus volume, the work given by Eq. (2-3a) is the area under the curve.

Thermal Energy Reservoir (TER). A thermal energy reservoir is a special kind of control mass that can undergo only heat interactions with its surround-

2-3 The Second Law of Thermodynamics, Entropy

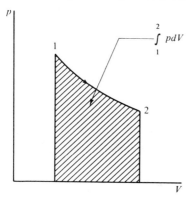

Figure 2-2 Area representation of work for an internally reversible expansion of a fluid in a piston–cylinder.

ings. Energy transferred into the TER in a heat interaction appears as an increase in its internal energy. However, its temperature is required to be uniform throughout and usually is assumed to be constant. Any change of state occurs in an internally reversible manner. The TER is always in an equilibrium state. A thermal energy reservoir is an idealization, but it can be approximated by large bodies of water (lakes, oceans), the Earth's atmosphere, a very large block of copper, and so on.

Further Corollaries of the Second Law. The following two corollaries of the second law play an important role in its practical utilization and follow by deduction from it. Each corollary refers to the thermal efficiency which is given for *any* heat engine by Eq. (2-2e).

Each also refers to a reversible heat engine. A reversible heat engine is one for which no irreversibilities are present within the device or its surroundings. Thus, the engine undergoes only internally reversible processes, and heat interactions with its surroundings occur only when the temperature difference between the engine working substance and the surroundings is vanishingly small.

> *Corollary 1.* The thermal efficiency of an actual (irreversible) heat engine is always less than the efficiency of a reversible heat engine when both are operating between the same two thermal energy reservoirs.
>
> *Corollary 2.* All reversible heat engines operating between the same two thermal energy reservoirs have the same thermal efficiency.

The proof for each of these statements utilizes the notion that if it were not valid a PMM2 would be possible.[3]

Kelvin Temperature Scale. Corollary 2 suggests that the thermal efficiency of a reversible heat engine operating between two thermal energy reservoirs depends only on the temperatures of the reservoirs and not on the nature of

---

[3]See, for example, Sec. 7.6 of Ref. [7], or Sec. 6-7 of Ref. [8].

the engine or its working substance. It follows that the ratio of the heat transfers is also related only to the temperatures

$$\frac{Q_R}{Q_A} = \phi(T_R, T_A)$$

where T_A is the temperature of the source on a scale to be defined and T_R is the temperature of the reservoir to which energy is rejected.

The *Kelvin temperature scale* is based on the choice $\phi(T_R, T_A) = T_R/T_A$. Then

$$\frac{Q_R}{Q_A} = \frac{T_R}{T_A} \tag{2-3b}$$

Equation (2-3b) defines only a ratio of temperatures. The specification of the Kelvin scale is completed by assigning a numerical value to one standard reference state. The state selected is the same used to define the gas scale (Sec. 2-1): at the triple state of water (see Fig. 2-4b) the temperature is specified to be 273.16 K. If a reversible heat engine is operated between a reservoir at the reference state and another reservoir at an unknown temperature T, then the latter temperature is related to the value at the reference state by

$$T = 273.16 \frac{Q}{Q'}$$

where Q is the energy received by the engine from the reservoir at temperature T and Q' is the energy rejected to the reservoir at the reference state. Thus, a temperature scale is defined which is valid over all ranges of temperature and which is independent of the thermometric substance. Over their common range of definition the Kelvin and gas scales are equivalent.

Carnot Efficiency. Consider the important special case of a heat engine operating between thermal energy reservoirs at temperatures T_A and T_R. If the engine is reversible, combination of Eqs. (2-2e) and (2-3b) results in

$$\eta_{carnot} = 1 - \frac{T_R}{T_A} \tag{2-3c}$$

This is called the *Carnot efficiency*. By invoking the two corollaries just stated, it should be evident that this is the efficiency for *all* reversible heat engines operating between two TERs at T_A and T_R. Moreover, it is the *maximum theoretical* efficiency that any heat engine, real or ideal, could have while operating between the same reservoirs.

Clausius Inequality. When a system passes through a complete cycle, a PMM2 is possible unless

2-3 The Second Law of Thermodynamics, Entropy

$$\oint \frac{\delta Q}{T} \leq 0 \tag{2-3d}$$

where δQ represents the energy received in a heat interaction at a part of the system boundary during a portion of the cycle and T is the corresponding absolute temperature at that part of the boundary.[4] The integral is to be performed over all parts of the boundary and over the entire cycle. The equality sign applies only for internally reversible operation. <u>The inequality sign applies in the presence of internal irreversibilities.</u> This relationship, known as the *Clausius inequality*, plays a central role in the formulation of the important property of matter known as entropy.

Defining Equation for Entropy. Entropy is introduced by showing that for a control mass $\int \delta Q/T$ has the same value between two states when the integral is evaluated for any internally reversible process between the states. This means that the value of the integral represents the difference in some property. The property is called entropy.

To elaborate on this reasoning, consider a control mass undergoing a cycle, shown in Fig. 2-3a, consisting of an internally reversible process along path A linking states 1 and 2 followed by an internally reversible process along path B between the same two states. By Eq. (2-3d)

$$\oint \frac{\delta Q}{T} = \int_1^2 \left(\frac{\delta Q}{T}\right)_A + \int_2^1 \left(\frac{\delta Q}{T}\right)_B = 0$$

This can be rearranged into

$$\int_1^2 \left(\frac{\delta Q}{T}\right)_A = \int_1^2 \left(\frac{\delta Q}{T}\right)_B$$

which shows that the integral of $\delta Q/T$ is the same for both paths. Since A and B are arbitrary, the integral of $\delta Q/T$ has the same value for any internally reversible process between the two states. Accordingly, for any internally

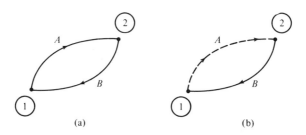

Figure 2-3 Control mass executing cycles.

---

[4]See, for example, Sec. 8-1 of Ref. [7], or Sec. 6-10 of Ref. [8].

reversible process, $\int_1^2 \delta Q/T$ defines the change in some property of the system between the two states. Selecting the symbol S to denote the property, the change is given by

$$S_2 - S_1 = \int_1^2 \left(\frac{\delta Q}{T}\right)_{\text{int. rev.}}$$

or

$$dS = \left(\frac{\delta Q}{T}\right)_{\text{int. rev.}} \tag{2-3e}$$

The property is named entropy. Entropy is an extensive property.

Increase of Entropy Principle. By application of the Clausius inequality, Eq. (2-3d), together with the definition of entropy, Eq. (2-3e), it is possible to obtain for a control mass

$$dS \geq \frac{\delta Q}{T} \tag{2-3f}$$

The equality sign applies only to internally reversible processes. The inequality applies for internally irreversible processes.

The inequality of Eq. (2-3f) is established by considering a control mass undergoing a cycle, shown in Fig. 2-3b, consisting of an internally irreversible process A followed by an internally reversible process B. The Clausius inequality for the cycle reads

$$\oint \frac{\delta Q}{T} = \int_1^2 \left(\frac{\delta Q}{T}\right)_A + \int_2^1 \left(\frac{\delta Q}{T}\right)_B < 0$$

Since B is internally reversible, the second integral can be evaluated with Eq. (2-3e). Thus

$$\int_1^2 \left(\frac{\delta Q}{T}\right)_A + (S_1 - S_2) < 0$$

or

$$(S_2 - S_1) > \int_1^2 \left(\frac{\delta Q}{T}\right)_A$$

Alternatively

$$dS > \left(\frac{\delta Q}{T}\right)_{\text{int. irrev.}}$$

Combining this with Eq. (2-3e) gives Eq. (2-3f).

In the absence of heat interactions with the surroundings Eq. (2-3f) reduces for any *control mass* to

2-3 The Second Law of Thermodynamics, Entropy

$$(dS)_{\substack{\text{adiabatic} \\ \text{control mass}}} \geq 0$$

That is, entropy increases when internal irreversibilities are present; entropy remains constant only in the limiting case of internally reversible operation; entropy cannot decrease. This is the increase of entropy principle.

According to one of the second law statements mentioned previously, systems left to themselves tend toward a state of equilibrium. The increase in entropy principle suggests that entropy must increase as that state is approached. The state of equilibrium is attained when entropy reaches its maximum possible value, consistent with the constraints on the system.

Entropy Production: Control Mass. Let a control mass receive energy in amount δQ from a TER. The entropy of the mass changes in accordance with

$$dS \geq \frac{\delta Q}{T} \tag{2-3f}$$

which shows that the increase in entropy of the control mass is greater when internal irreversibilities are present than when they are not. Thus it may be said that when there are internal irreversibilities entropy is *produced* within the control mass. The amount of entropy produced, $\delta\sigma$, within the control mass is calculated as follows

$$\delta\sigma \equiv dS - \frac{\delta Q}{T} \tag{a}$$

Since the entropy production depends on the kind of process and not solely on the end states of the process, it is not a property.

The following can be stated about the entropy production:

$\delta\sigma > 0$ internally irreversible process
$\delta\sigma = 0$ internally reversible process
$\delta\sigma < 0$ impossible process

The TER supplying the energy undergoes only internally reversible processes. It experiences a decrease in entropy

$$dS_{\text{TER}} = -\frac{\delta Q}{T}$$

The minus sign signals that the energy flow is out of the TER. Also, in writing this it is assumed the temperature difference between the TER and the point on the boundary of the system where energy is received is vanishingly small. Since a heat transfer of energy from the TER to the control mass results in a decrease in entropy for the TER and an increase in entropy for the control mass, it is possible to think of entropy *flowing* between the two as a result of their interaction. It is then natural to identify the term $\delta Q/T$ as *the flow of entropy associated with the heat interaction.*

The main points of this discussion can be summarized as follows. When there is a heat transfer of energy δQ to a control mass, there is a flow of entropy to it in amount $\delta Q/T$, where T is the temperature at the point of the control surface where the energy enters. If the energy flow is out of the system, then entropy exits in amount $\delta Q/T$. If there is a net inflow of entropy with heat, the entropy of the control mass increases by at least this amount, as shown by Eq. (2-3f). The amount by which the actual entropy increase of the control mass exceeds that due to the net entropy inflow with heat represents the entropy produced within the control mass as a result of internal irreversibilities.[5]

It is convenient to have Eq. (a) in the form

$$dS = \frac{\delta Q}{T} + \delta\sigma \qquad (2\text{-}3g)$$

which for a process of the control mass becomes

$$\Delta S = \int \frac{\delta Q}{T} + \sigma \qquad (2\text{-}3h)$$

Since the entropy production cannot be less than zero, these equations show that the entropy of a control mass can decrease only when energy is removed in a heat interaction.

Example 2-2. Consider a rod through which there is a steady heat transfer of energy. The ends of the rod are held at constant temperatures T_1 and T_2, it is insulated along its length, and its state does not change with time. Determine the entropy carried in and carried out with heat, and determine the entropy production. The situation is shown in Fig. E2-2.

Solution. Letting Q denote the energy flow into the rod by heat transfer during a specified time interval, an energy equation for the rod at steady-state shows that the energy out by heat transfer is $(-Q)$.

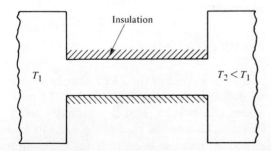

Figure E2-2

---

[5] Note that there is *no* flow of entropy associated with a work interaction.

The flow of entropy into the rod associated with the heat interaction is Q/T_1, and the flow out is $(-Q)/T_2$. Applying Eq. (2-3h)

$$0 = \frac{Q}{T_1} - \frac{Q}{T_2} + \sigma$$

where the change in entropy for the rod, ΔS, is set to zero since its state does not change with time.

Solving for the entropy production, σ,

$$\sigma = Q\left(\frac{1}{T_2} - \frac{1}{T_1}\right) = \frac{Q}{T_1 T_2}(T_1 - T_2)$$

Since $T_1 > T_2$, the entropy production is positive. The entropy production in this case is directly chargeable to the internal irreversibility that exists: heat transfer through a finite temperature difference (see Example 2-1). The entropy production approaches zero as the two temperatures approach each other, that is, as the heat transfer approaches reversibility.

Entropy Production: Control Volume. Using an approach similar to that employed in Sec. 2-2 to develop the control volume energy equation, an equation accounting for the entropy production in a control volume is derived as follows.

Figure 2-1 shows a control mass which at time t is the sum of the mass within the control volume and the mass δm within the shaded area adjacent to section i. During a time interval dt the mass δm crosses i to join the material initially within the control volume. Heat transfers are permitted at various points on the boundary and work interactions may also occur.

At time t the entropy of the control mass equals the entropy in the control volume plus the entropy of the shaded portion. The entropy of the shaded portion is $s_i \delta m$, where s_i is the specific entropy (entropy per unit mass).

$$S_{cm}(t) = S_{cv}(t) + s_i \delta m \tag{b}$$

At $t + dt$ the boundary of the control mass coincides with that of the control volume and so

$$S_{cm}(t + dt) = S_{cv}(t + dt) \tag{c}$$

The entropy production for the control mass is

$$\delta\sigma = [S_{cm}(t + dt) - S_{cm}(t)] - \sum_j \frac{\delta Q_j}{T_j}$$

T_j is the temperature at the point where δQ_j enters. Upon inserting Eqs. (b) and (c)

$$\delta\sigma = dS_{cv} - s_i \delta m - \sum_j \frac{\delta Q_j}{T_j} \tag{d}$$

where $dS_{cv} = S_{cv}(t + dt) - S_{cv}(t)$.

Noting that Eq. (d) is written in terms of quantities related to the control volume, it can be viewed as an expression for the entropy production within the control volume. This definition of entropy production can readily be extended to a control volume with several flow inlets and exits. When expressed on a time rate basis

$$\dot{\sigma} = \frac{dS_{cv}}{dt} + \sum_e \dot{m}_e s_e - \sum_i \dot{m}_i s_i - \sum_j \frac{\dot{Q}_j}{T_j}$$

Or, in what is frequently a convenient form

$$\frac{dS_{cv}}{dt} = \sum_j \frac{\dot{Q}_j}{T_j} + \sum_i \dot{m}_i s_i - \sum_e \dot{m}_e s_e + \dot{\sigma} \qquad (2\text{-}3i)$$

In this equation, $\dot{\sigma}$ is the rate of entropy production. The second law requires that $\dot{\sigma} \geq 0$, where the equality applies only when no internal irreversibilities are present within the control volume. The inequality applies when internal irreversibilities are present. The terms $\dot{m}_i s_i$ and $\dot{m}_e s_e$ account for, respectively, entropy transfers associated with mass flow into and out of the control volume. The terms $\dot{Q}_j/T_j$ represent rates of entropy transfer associated with the heat interactions.

A further generalization is

$$\frac{dS_{cv}}{dt} = \int_\alpha (q_s/T_s) d\alpha + \sum_i \dot{m}_i s_i - \sum_e \dot{m}_e s_e + \dot{\sigma} \qquad (2\text{-}3j)$$

where T_s is the temperature at the point on the control surface where the heat flux (heat transfer rate per unit of area) is q_s. The integral is performed over the surface area of the control volume.

2-4 Simple Compressible Substances

A wide range of industrially important gases and liquids are satisfactorily modeled as *simple compressible substances*. A simple compressible substance is one for which the only reversible work mode is volume change ($\int pdV$ work).

Although an indefinite number of thermostatic properties can be associated with a state, the values of these properties are not all independent. For a simple compressible substance of fixed composition, the number of independent thermostatic properties is two. This is known as the *state principle* for simple compressible substances.

This section presents relationships among the thermostatic properties of simple compressible substances of fixed composition.

Basic Property Relations. Properties such as internal energy, enthalpy, and entropy cannot be directly measured. These properties are calculated from

2-4 Simple Compressible Substances

other property data. The calculations use the *TdS equations* and relations that can be derived from them.

The energy equation in differential form for a control mass at rest is

$$dU = \delta Q - \delta W$$

For an internally reversible process, $\delta Q = TdS$, Eq. (2-3e), and $\delta W = pdV$, Eq. (2-3a). Thus

$$dU = TdS - pdV \tag{2-4a}$$

This is the *first TdS* equation for simple compressible substances of fixed composition.

Equation (2-4a) holds for any change in state regardless of the nature of the process, because it is a relation between properties and changes in values of properties. However, it is only for internally reversible processes that TdS is a heat transfer of energy and pdV is a work transfer of energy.

Since $H = U + pV$, $dH = dU + pdV + Vdp$. Combining this with Eq. (2-4a), the *second TdS equation* follows

$$dH = TdS + Vdp \tag{2-4b}$$

Using the same approach, a further relation can be derived from the *Gibbs free energy*, $G = H - TS$. The result is

$$dG = Vdp - SdT \tag{2-4c}$$

The last three equations can be expressed on a unit mass (or a per mole) basis.

$$\begin{aligned} du &= Tds - pdv \\ dh &= Tds + vdp \\ dg &= vdp - sdT \end{aligned} \tag{2-4d}$$

The properties c_v and c_p are defined as follows:

$$\begin{aligned} c_v &= \left(\frac{\partial u}{\partial T}\right)_v \\ c_p &= \left(\frac{\partial h}{\partial T}\right)_p \end{aligned} \tag{2-4e}$$

where the subscripts v and p denote, respectively, the variables held fixed in the differentiation process. These properties are known commonly as *specific heats*. The property k is simply the ratio

$$k = c_p/c_v$$

p-v-T Relations. The thermostatic properties pressure, specific volume, and temperature are relatively easily measured and considerable *p-v-T* data have been accumulated for pure substances such as water, nitrogen, and so on. According to the state principle, any one of the three properties *p-v-T* can be determined as a function of the remaining two. For example, $p = f(v, T)$. Any such relation between these properties is called an *equation of state*. Equations of state can be expressed in tabular, graphical, and analytical forms.

The graph of a function $p = f(v, T)$ is a surface in three-dimensional space. This surface is complicated in appearance. To illustrate, Fig. 2-4a shows

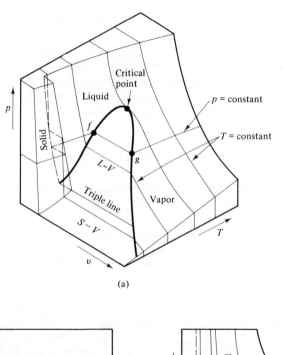

(a)

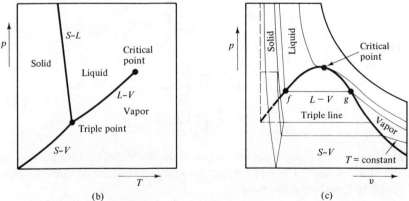

Figure 2-4 Pressure–specific volume–temperature surface and projections for water.

2-4 Simple Compressible Substances

the p-v-T relationship for water, a substance which expands upon freezing. Figure 2-4b shows the projection of the surface onto the pressure–temperature plane. The projection onto the p–v plane is shown in Fig. 2-4c.

A quantity of matter that is homogeneous in chemical composition and physical structure is called a *phase*. To be homogeneous in physical structure the matter is all vapor (gas),[6] all liquid, or all solid. Figure 2-4 has three regions labeled solid, liquid, and vapor where the substance exists only in a single phase. Between the single-phase regions lie *two-phase* regions, where two phases coexist in equilibrium. The line separating a single-phase region from a two-phase region is called a *saturation line*. Any state represented by a point on a saturation line is a *saturation state*. Thus, the line separating the liquid phase and the two-phase liquid–vapor region is the saturated liquid line. The state denoted by f on Fig. 2-4c is a saturated liquid state. The saturated vapor line separates the vapor region and the two-phase liquid–vapor region. The state denoted by g on Fig. 2-4c is a saturated vapor state.

As shown in Fig. 2-4, the saturated liquid line and the saturated vapor line meet at a point, called the *critical point*. The pressure at the critical point is called the *critical pressure* p_c, and the temperature at the critical point is called the *critical temperature* T_c. Critical point data for a number of substances are given in Appendix Table B-1.

Whenever a phase change occurs during constant pressure heating or cooling, the temperature remains constant as long as both phases are present. Accordingly, in the two-phase liquid–vapor region, a line of constant pressure is also a line of constant temperature. For a specified pressure, the corresponding temperature is called the *saturation temperature*. For a specified temperature, the corresponding pressure is called the *saturation pressure*. The region to the right of the saturated vapor line is often referred to as the *superheated vapor region* because the vapor exists at a temperature greater than the saturation temperature for its pressure. The region to the left of the saturated liquid line is also known as the *compressed liquid region* because the liquid is at a pressure higher than the saturation pressure for its temperature.

When a mixture of liquid and vapor coexist in equilibrium, the liquid phase is at the saturated liquid state and the vapor is at the corresponding saturated vapor state. The total volume of any such mixture is $V = V_f + V_g$, where f denotes saturated liquid and g saturated vapor. This may be written as

$$mv = m_f v_f + m_g v_g$$

where m and v denote mass and specific volume, respectively. Dividing by the total mass of the mixture m and letting the mass fraction of the vapor in the mixture, m_g/m, be symbolized by x, called the *quality*, the specific volume v of the mixture is

---

[6] *Vapor* and *gas* are used interchangeably here.

$$v = (1-x)v_f + xv_g$$
$$= v_f + xv_{fg}$$

where v_{fg} is defined as $v_g - v_f$. Expressions analogous to these can be derived for internal energy, enthalpy, and entropy.

The general pattern of the p-v-T relation can also be conveniently shown for liquid and vapor states in terms of the *compressibility* factor Z. By definition

$$Z = \frac{p\bar{v}}{\bar{R}T}$$

where $\bar{R}$ is the universal gas constant (Sec. 2-1).

Using the compressibility factor, *generalized* charts can be developed from which reasonable approximations to the p-v-T behavior of many substances can be obtained. In one form of generalized chart the compressibility factor Z is plotted versus the *reduced pressure* p_r and *reduced temperature* T_r, defined by

$$p_r = \frac{p}{p_c} \qquad T_r = \frac{T}{T_c}$$

Figure 2-5 shows the experimental data for 10 different gases on a generalized compressibility chart which has Z, p_r, and T_r as the coordinates. The data correlate with an overall average deviation of about 1%. Accordingly, the chart can be used with acceptable accuracy for a wide range of engineering applications. Appendix Figure B-1 is a generalized compressibility chart which is more suitable for engineering calculations.

Property Tables for Gases and Liquids. The discussion so far has centered on graphical pressure-specific volume–temperature relations. Tabular presentations are also available for many substances. The tables normally include other properties useful for thermodynamic analyses, such as internal energy, enthalpy, and entropy.

Values for u, h, and s are calculated for real gases and liquids from p-v-T relations, specific heats, and other thermodynamic data. The results of such calculations, supplemented by data taken from direct physical measurement or determined from these measurements, are presented in tabular form for ease of use. Appendix Tables B-2 through B-4 are examples. It is assumed that the form of the tables and the way they are used is familiar.

In using Appendix Tables B-2 through B-4, it should be recognized that the enthalpy, internal energy, and entropy values appearing in them are calculated relative to arbitrary datums and that the datum used varies from substance to substance. For example, the internal energy of saturated liquid water at 32.02°F is set to zero (Appendix Table B-2A), but for Refrigerant 12 the saturated liquid enthalpy is zero at −40°F (Appendix Table B-3A). The use of values for a particular property determined relative to an arbitrary datum is satis-

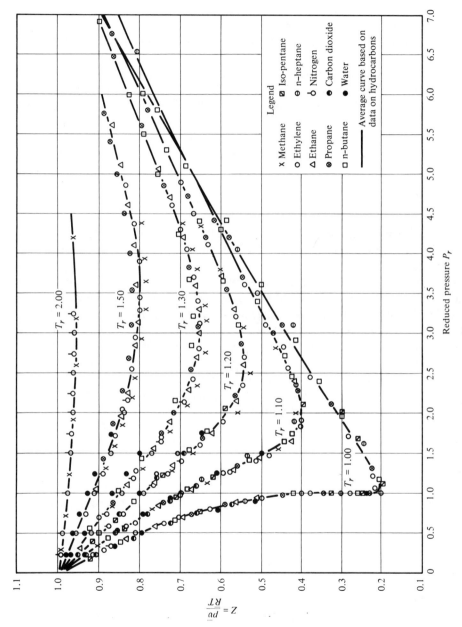

Figure 2-5 A generalized compressibility chart for various gases. [Reprinted with permission from G. Su, "Modified Law of Corresponding States," *Ind. Eng. Chem.*, **38**, August 1946, 803–806. Copyright 1946 American Chemical Society.]

factory when the calculations being performed involve only differences in that property, for then the datum cancels. Care must be taken, however, when there are changes in composition during the process. The approach to be followed when composition changes due to chemical reaction is explained in Sec. 7-2.

Ideal Gas Property Relations. By inspection of a generalized compressibility chart it can be concluded that when p_r is small, and for many states when T_r is large, the value of the compressibility factor Z is closely 1. Stated in other words, for pressures that are low relative to p_c, and for many states with temperatures high relative to T_c, the compressibility factor approaches a value of 1. Within the indicated limits, it may be assumed with reasonable accuracy that $Z = 1$. That is

$$p\bar{v} = \bar{R}T \qquad (2\text{-}4\text{f})$$

It can be shown that the internal energy and enthalpy for any gas whose equation of state is *exactly* given by Eq. (2-4f) can depend only on temperature (see Problem 2-7).

These considerations lead to the introduction of an *ideal gas model* for each real gas. For example, the ideal gas model of nitrogen obeys the equation of state $p\bar{v} = \bar{R}T$ and its internal energy is a function of temperature alone. The ideal gas model of oxygen also obeys $p\bar{v} = \bar{R}T$ and its internal energy is also a function of temperature alone, but the internal energy function is not the same as for nitrogen since each gas has a unique internal structure. The model exhibits the same molecular weight as for the corresponding real gas. The real gas approaches the model in the limit of low pressure. At other states the real behavior may deviate substantially from the predictions of the model.

Dividing the ideal gas equation of state by the molecular weight M, the equation is placed on a unit mass basis

$$pv = RT$$

where R, defined as $\bar{R}/M$, is the *specific* gas constant. Other forms in common use are

$$pV = N\bar{R}T$$
$$pV = mRT$$

Frequently used values of $\bar{R}$ are

$$\bar{R} = 8.315 \frac{\text{kJ}}{(\text{kgmol})(\text{K})}$$
$$= 1545 \frac{\text{ft lbf}}{(\text{lbmol})(°\text{R})}$$

2-4 Simple Compressible Substances

Since $p\bar{v} = \bar{R}T$ for an ideal gas, the ideal gas enthalpy and internal energy are related by $\bar{h}(T) = \bar{u}(T) + \bar{R}T$. Differentiating with respect to temperature

$$\frac{d\bar{h}}{dT} = \frac{d\bar{u}}{dT} + \bar{R}$$

and introducing the specific heat concept, Eqs. (2-4e),

$$\bar{c}_p(T) = \bar{c}_v(T) + \bar{R} \tag{2-4g}$$

The two ideal gas specific heats depend on temperature alone, and their difference is the gas constant. Since $k = c_p/c_v$

$$\bar{c}_p = \frac{k\bar{R}}{k-1} \qquad \bar{c}_v = \frac{\bar{R}}{k-1} \tag{2-4h}$$

(see Problem 2-1).

Specific heat data can be obtained by direct measurement. When extrapolated to zero pressure, ideal gas model specific heats result. Ideal gas specific heats can also be obtained from theory based on molecular models of matter using spectroscopic measurements. Ideal gas specific heat functions for a number of substances are reported in Appendix Table B-5.

Since $\bar{c}_v = d\bar{u}/dT$ and $\bar{c}_p = d\bar{h}/dT$ for an ideal gas, expressions for internal energy and enthalpy can be obtained by integration

$$\bar{u}(T) = \int_{T_R}^{T} \bar{c}_v(T)dT + \bar{u}(T_R)$$

$$\bar{h}(T) = \int_{T_R}^{T} \bar{c}_p(T)dT + \bar{h}(T_R) \tag{2-4i}$$

where T_R is an arbitrary reference temperature.

The first TdS equation, $Td\bar{s} = d\bar{u} + pd\bar{v}$, can be used to determine the entropy change of an ideal gas between two states. Using $d\bar{u} = \bar{c}_v dT$ and $p/T = \bar{R}/\bar{v}$

$$d\bar{s} = \frac{\bar{c}_v(T)dT}{T} + \frac{\bar{R}}{\bar{v}} d\bar{v}$$

Integrating

$$\bar{s}(T_2, v_2) - \bar{s}(T_1, v_1) = \int_{T_1}^{T_2} \frac{\bar{c}_v(T)}{T} dT + \bar{R}\ln\left(\frac{v_2}{v_1}\right) \tag{2-4j}$$

Similarly, from the second TdS equation follows

$$\bar{s}(T_2, p_2) - \bar{s}(T_1, p_1) = \int_{T_1}^{T_2} \frac{\bar{c}_p(T)}{T} dT - \bar{R}\ln\left(\frac{p_2}{p_1}\right) \tag{2-4k}$$

(see Problem 2-2). This can be rewritten using the definition

$$\bar{s}'(T) \equiv \int_{T_R}^{T} \frac{\bar{c}_p(T)}{T} \, dT \tag{a}$$

as

$$\bar{s}(T_2, p_2) - \bar{s}(T_1, p_1) = \bar{s}'(T_2) - \bar{s}'(T_1) - \bar{R} \ln(p_2/p_1) \tag{2-4l}$$

Reference to Eqs. (2-4i) and Eq. (a) shows that upon specifying the reference temperature T_R the quantities $\bar{u}(T)$, $\bar{h}(T)$, $\bar{s}'(T)$ can be tabulated versus temperature. Appendix Table B-6 gives such tabulations for each of several ideal gases based on the specification $\bar{h} = \bar{u} = \bar{s}' = 0$ at $T_R = 0$ K.

When the temperature interval is relatively small, the ideal gas specific heats are nearly constant. It is often convenient in such instances to assume them to be constant, usually the arithmetic average over the interval. The expressions for changes in internal energy, enthalpy, and entropy then appear as

$$\bar{u}(T_2) - \bar{u}(T_1) = \bar{c}_v[T_2 - T_1]$$
$$\bar{h}(T_2) - \bar{h}(T_1) = \bar{c}_p[T_2 - T_1]$$
$$\bar{s}(T_2, v_2) - \bar{s}(T_1, v_1) = \bar{c}_v \ln(T_2/T_1) + \bar{R} \ln(\bar{v}_2/\bar{v}_1)$$
$$\bar{s}(T_2, p_2) - \bar{s}(T_1, p_1) = \bar{c}_p \ln(T_2/T_1) - \bar{R} \ln(p_2/p_1)$$

Ideal Gas Mixtures. Under consideration in this section is a phase consisting of a mixture of n gases. It is assumed that each gas is uninfluenced by the presence of the others, each can be treated as an ideal gas, and each acts as if it exists separately at the volume and temperature of the mixture. This is the *Dalton* mixture model.

Writing the ideal gas equation of state for the mixture as a whole and for any component k

$$pV = N\bar{R}T$$
$$p_k V = N_k \bar{R}T$$

where p is the mixture pressure and p_k is the *partial pressure* of component k. The partial pressure p_k is the pressure gas k would exert if N_k moles occupied the volume alone at the mixture temperature.

Forming a ratio of these two equations

$$p_k = \frac{N_k}{N} p = x_k p \tag{2-4m}$$

where x_k, defined as N_k/N, is the *mole fraction* of component k. Since $\sum_{k=1}^{n} x_k = 1$, it follows that

$$p = \sum_{k=1}^{n} p_k$$

2-4 Simple Compressible Substances

The pressure of a mixture of ideal gases is equal to the sum of the partial pressures of the components. This is known as *Dalton's law of additive partial pressures*.

The internal energy, enthalpy, and entropy of the mixture can be determined as the sum of the respective properties of the component gases, provided that the contribution from each component is evaluated as if it exists alone at the temperature and volume of the mixture. Thus

$$U = \sum_{k=1}^{n} N_k \bar{u}_k \quad \text{or} \quad \bar{u} = \sum_{k=1}^{n} x_k \bar{u}_k$$

$$H = \sum_{k=1}^{n} N_k \bar{h}_k \quad \text{or} \quad \bar{h} = \sum_{k=1}^{n} x_k \bar{h}_k \qquad (2\text{-}4n)$$

$$S = \sum_{k=1}^{n} N_k \bar{s}_k \quad \text{or} \quad \bar{s} = \sum_{k=1}^{n} x_k \bar{s}_k$$

Since the internal energy and enthalpy for an ideal gas depend only on temperature, the $\bar{u}_k$ and $\bar{h}_k$ terms appearing in these equations are evaluated at the temperature of the mixture. Entropy is a function of *two* independent properties even for an ideal gas. Accordingly, the $\bar{s}_k$ terms are evaluated either at the temperature and volume of the mixture, or at the mixture temperature and the partial pressure p_k of the component. In the latter case

$$\begin{aligned} S &= \sum_{k=1}^{n} N_k \bar{s}_k(T, p_k) \\ &= \sum_{k=1}^{n} N_k \bar{s}_k(T, x_k p) \end{aligned} \qquad (2\text{-}4o)$$

Differentiation of $\bar{u} = \sum_{k=1}^{n} x_k \bar{u}_k$ and $\bar{h} = \sum_{k=1}^{n} x_k \bar{h}_k$ with respect to temperature results, respectively, in expressions for the two specific heats $\bar{c}_v$ and $\bar{c}_p$ for the mixture in terms of the corresponding specific heats of the components.

$$\begin{aligned} \bar{c}_v &= \sum_{k=1}^{n} x_k \bar{c}_{vk} \\ \bar{c}_p &= \sum_{k=1}^{n} x_k \bar{c}_{pk} \end{aligned} \qquad (2\text{-}4p)$$

where $\bar{c}_{vk} = d\bar{u}_k/dT$, $\bar{c}_{pk} = d\bar{h}_k/dT$.

The molecular weight M of the mixture is determined in terms of the molecular weights M_k of the components as

$$M = \sum_{k=1}^{n} x_k M_k$$

Property Relations for Incompressible Liquids. Reference to Appendix Table B-2D shows that for the liquid phase of water there are regions where the

exact differential (see Problems 2-3 and 2-4). Letting $M(p, T) = v$ and $N(p, T) = -s$, apply the test for exactness to obtain

$$\left(\frac{\partial v}{\partial T}\right)_p = -\left(\frac{\partial s}{\partial p}\right)_T$$

This is known as a *Maxwell* relation. It permits the quantity $\partial s/\partial p)_T$ to be eliminated in favor of $\partial v/\partial T)_p$ which can be readily evaluated from experimental p-v-T data.

2-6 Using the Maxwell relation

$$\left(\frac{\partial v}{\partial T}\right)_p = -\left(\frac{\partial s}{\partial p}\right)_T$$

(see Problem 2-5) and the second TdS equation, show that

$$\left(\frac{\partial h}{\partial p}\right)_T = v - T\left(\frac{\partial v}{\partial T}\right)_p$$

2-7 Using the result of Problem 2-6, show that for any gas whose equation of state is given *exactly* by $pv = RT$ the enthalpy is independent of pressure and must depend only on temperature. Then show that its internal energy also must depend only on temperature.

2-8 Consider a control mass consisting of an ideal gas undergoing a process in which there is no change in entropy. Assuming the specific heats are constant for the process, show that the pressure and specific volume at the two end states are related by

$$p_2 v_2^k = p_1 v_1^k$$

where k is the ratio of the specific heats.

2-9 As shown in Fig. P2-9, N_1 moles of gas 1 at pressure p and temperature T are mixed with N_2 moles of gas 2 also at p and T in an adiabatic process. The final mixture pressure is p. Assuming ideal gases, show that the final mixture temperature is T. Also show that the entropy production in the mixing process is

$$\sigma = -\bar{R}(N_1 \ln x_1 + N_2 \ln x_2)$$

where $x_i = N_i/(N_1 + N_2)$, $i = 1, 2$.

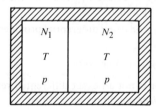 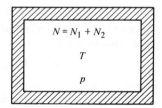

Figure P2-9

2-10 N moles of an ideal gas expand isothermally in a piston–cylinder device from V_1 to V_2. Show that the work of the internally reversible process is

$$W = N\bar{R}T \ln (V_2/V_1)$$

Evaluate the heat transfer Q for the process.

Problems

2-11 Water (assume incompressible) flows into an initially empty cylindrical tank at a volumetric flow rate of 20 ft³/min. The tank diameter is 10 ft. Simultaneously, water is drained at a volumetric flow rate proportional to the height of liquid in the tank: $2h$ ft³/min, where h is the height of liquid. Determine the height as a function of time, $h(t)$.

2-12 A certain power cycle consists of four processes in series
  1–2: Heat transfer *to* the working fluid in amount 100 Btu. No shaft work.
  2–3: Adiabatic. Shaft work out.
  3–4: Heat transfer *from* the working fluid in amount 60 Btu. No shaft work.
  4–1: Adiabatic. Shaft work done on the fluid in amount 10 Btu.
Determine the net work of the cycle, the work of process 2–3, and the thermal efficiency.

2-13 An inventor claims to have developed an engine which at steady-state delivers shaft power at a rate of 500 Btu/s for a heat addition of 500 Btu/s at 540°F. There is no heat rejection. Evaluate this claim quantitatively using both the first and second laws of thermodynamics.

2-14 A gear box, in steady-state operation at 110° F, receives 20 horsepower (hp) along a shaft while delivering 18 hp along the output shaft. Determine the rate of entropy production within the gear box.

2-15 A certain amount of air is initially at 20 lbf/in.² and 140° F. Can a final state defined by 15 lbf/in.² and 60° F be attained in an adiabatic process? Assume the ideal gas model.

2-16 One lb of air is contained in a closed, rigid, insulated tank. Initially the temperature is 40°F and the pressure is one atmosphere. Paddlewheel work is done on the air in amount 85 Btu. Assuming the ideal gas model, determine the final temperature and pressure, and the amount of entropy produced in the process.

2-17 Consider a closed, rigid tank containing a fluid at steady-state. The fluid is at a temperature of 140°F and is stirred continuously by a paddlewheel. If the rate energy enters due to the stirring is 0.3 Btu/s, determine the heat transfer rate and the rate of entropy production.

2-18 The following steady-state operating data are claimed for a steam turbine. At the inlet, the steam is at 100 lbf/in.² and 500°F. At the exit the pressure is 2 lbf/in.². The flow rate is 30,000 lb/hr and the device develops 3800 hp in output shaft work. Assuming adiabatic operation and neglecting kinetic and potential energy changes, evaluate this claim.

2-19 Water vapor flows steadily through a well insulated nozzle. At the inlet the pressure is 100 lbf/in.² and the temperature is 600°F. At the exit the pressure is 30 lbf/in.². Neglecting the potential energy change and the inlet kinetic energy, determine the theoretical maximum exit velocity.

2-20 Consider a compressor using Refrigerant 12 as the working substance. At the inlet, the pressure is 20 lbf/in.² and the temperature is 40°F. At the exit, the pressure is 120 lbf/in.². Assuming the device operates at steady-state and adiabatically, determine the theoretical minimum shaft work required per unit of mass flowing through. Neglect kinetic and potential energy changes.

2-21 Steam at a pressure of 100 lbf/in.² and 450°F is contained in a vast reservoir. Connected to the reservoir by a valve is a small initially evacuated tank having a volume of 1 ft³. The valve is opened and the tank is rapidly filled until the tank pressure is 100 lbf/in.². The valve is then closed. Assume the filling process is essentially adiabatic, kinetic and potential energy changes are negligible, and the state of the reservoir remains constant. Determine the final temperature of the steam within the tank. Also determine the amount of entropy produced within the tank. Hint: Take the state of the steam entering the tank to be the same as that of the reservoir.

2-22 Consider the two-dimensional stationary control volume shown in Fig. P2-22. The fluid flowing through it is assumed to be incompressible.

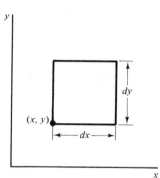

Figure P2-22

(a) Apply the principle of conservation of mass to obtain the *continuity* equation

$$0 = \frac{\partial \mathcal{V}_x}{\partial x} + \frac{\partial \mathcal{V}_y}{\partial y} \tag{a}$$

where $\mathcal{V}_x$ and $\mathcal{V}_y$ denote, respectively, the x and y components of the fluid velocity vector.

(b) Apply the control volume form of the entropy equation to obtain the equation of change for entropy

$$\rho \frac{\partial s}{\partial t} = -\left[\frac{\partial}{\partial x}\left(\frac{q_x}{T}\right) + \frac{\partial}{\partial y}\left(\frac{q_y}{T}\right)\right] - \rho\left[\frac{\partial(\mathcal{V}_x s)}{\partial x} + \frac{\partial(\mathcal{V}_y s)}{\partial y}\right] + \sigma \tag{b}$$

where q_x and q_y denote, respectively, the x and y components of the heat flux vector and σ is the time rate of entropy production *per unit of volume*. Use Eq. (a) to show Eq. (b) can be expressed in the form

$$\rho\left(\frac{\partial s}{\partial t} + \mathcal{V}_x \frac{\partial s}{\partial x} + \mathcal{V}_y \frac{\partial s}{\partial y}\right) = -\left[\frac{\partial}{\partial x}\left(\frac{q_x}{T}\right) + \frac{\partial}{\partial y}\left(\frac{q_y}{T}\right)\right] + \sigma \tag{c}$$

(c) Referring to the literature, confirm that the equation of change for internal energy can be written as

$$\rho\left(\frac{\partial u}{\partial t} + \mathcal{V}_x \frac{\partial u}{\partial x} + \mathcal{V}_y \frac{\partial u}{\partial y}\right) = -\left(\frac{\partial q_x}{\partial x} + \frac{\partial q_y}{\partial y}\right) + \phi_v \tag{d}$$

where ϕ_v accounts for the irreversible conversion of *mechanical* energy to *internal* energy

$$\phi_v = \mu\left\{2\left[\left(\frac{\partial \mathcal{V}_x}{\partial x}\right)^2 + \left(\frac{\partial \mathcal{V}_y}{\partial y}\right)^2\right] + \left[\frac{\partial \mathcal{V}_y}{\partial x} + \frac{\partial \mathcal{V}_x}{\partial y}\right]^2\right\} \tag{e}$$

where μ is the fluid viscosity (see for example R. B. Bird, W. E. Stewart, and E. N. Lightfoot, *Transport Phenomena*, John Wiley & Sons, New York, 1960, Secs. 3-3, and 10-1).

(d) Noting that for an incompressible fluid $du = Tds$, combine Eqs. (c) and (d) to obtain

$$\sigma = -\frac{1}{T^2}\left(q_x \frac{\partial T}{\partial x} + q_y \frac{\partial T}{\partial y}\right) + \frac{\phi_v}{T} \tag{f}$$

(e) Use *Fourier's* conduction model

$$q_x = -k\frac{\partial T}{\partial x} \qquad q_y = -k\frac{\partial T}{\partial y}$$

where k is the fluid thermal conductivity, and Eq. (e) to rewrite Eq. (f) as

$$\sigma = \frac{k}{T^2}\left[\left(\frac{\partial T}{\partial x}\right)^2 + \left(\frac{\partial T}{\partial y}\right)^2\right] + \frac{\mu}{T}\left\{2\left[\left(\frac{\partial v_x}{\partial x}\right)^2 + \left(\frac{\partial v_y}{\partial y}\right)^2\right] + \left[\frac{\partial v_y}{\partial x} + \frac{\partial v_x}{\partial y}\right]^2\right\} \tag{g}$$

Eqs. (f) and (g) show that the entropy production is determined by two additive contributions. The first is associated with internal heat transfer and the second with fluid friction.

Chapter 3

Thermomechanical Availability

This chapter begins the development of the availability concept. As part of the presentation availability equations are introduced for a control mass and for a control volume. A number of solved examples are included to illustrate the availability concept as well as the use of the availability equations.

3-1 Introduction

Section 1-2 presents an illustration that distinguishes the notion of energy *quantity* from another important aspect of energy, its *potential* or *quality*. As another example, it is easy to accept that one unit of energy exiting a power plant in a low temperature stream of cooling water does not have the same potential as one unit exiting as electricity. A precise formulation of these ideas is achieved through the property availability which takes into account both the quantity and potential of energy.

3-2 The Environment

Any system of interest, whether a component in a larger system such as a turbine in a power plant or the larger system itself, operates within an environment of some kind. The sections to follow bring out the idea that the numerical value for availability depends in part on the make-up of the environment.

It is important to distinguish between the environment and the system's surroundings. The surroundings comprise everything not included in the system. One part of the surroundings is some portion of the Earth and its atmosphere, the intensive properties of which do not change significantly as a result of any of the processes under consideration. It is to this that the term environment applies.

The physical world is complicated, and to include every detail in an analysis is not practical. Accordingly, in describing the environment simplifications are made. The result is a model. The validity and utility of an analysis making use of any model is of course restricted by the idealizations made in formulating it. At several locations in this book, models for environments appropriate to particular kinds of applications are described. A few general characteristics are now stated.

The environment is assumed to be large in extent and homogeneous in temperature and pressure. All parts are at rest relative to one another. It is a source (or sink) of internal energy which can be freely drawn upon (or added to) without change in its intensive properties. It experiences only internally reversible processes in which the sole work mode is associated with volume change (pdV work). It receives[1] heat transfers of energy at the uniform temperature T_o.

In the present chapter the composition of the environment is considered to be fixed. Changes in the extensive properties, internal energy, entropy, and volume of the environment are related through the *first TdS* equation, Eq. (2-4a)

$$\Delta U^o = T_o \Delta S^o - p_o \Delta V^o \tag{3-2a}$$

In this expression, U^o, S^o, and V^o denote, respectively, the internal energy, entropy, and volume of the environment, T_o is the temperature of the environment and p_o is its pressure.

In keeping with the idea that the environment has to do with the actual physical world, the values for p_o and T_o used throughout a particular analysis are normally taken as *typical* environmental conditions, such as 1 atm and 77°F (25°C). In Chaps. 6 and 7, it will be necessary also to describe the chemical make-up of the environment, and take into account changes in its composition.

3-3 The Restricted Dead State

A case of particular importance in the formulation of the availability concept is when a control mass passes from some specified state to one in which it is in thermal and mechanical equilibrium (Sec. 2-1) with the environment. At this state there can be no spontaneous change within the control mass or within the environment, nor can there be a spontaneous interaction between the two.

---

[1] The term "receives" also includes the possibility that energy can be drawn from the environment by heat transfer.

This state of the control mass within the environment can be called a *dead state*.

The dead state may be simply conceptualized: The control mass contents, internally in equilibrium, are imagined to be sealed in a flexible envelope impervious to mass flow, at rest with the environment, and in temperature and pressure equilibrium with it. At the dead state, the control mass is at temperature T_o and pressure p_o.

The contents of the control mass are not permitted to mix with the environment or enter into chemical reaction with environmental components. Effects such as these are considered in Chaps. 6 and 7, requiring a more detailed formulation of the environment and dead state concepts than considered here. As a way of indicating this limitation, the dead state in use through Chap. 5 is referred to as the *restricted* dead state.

3-4 Fundamentals

Introduction. This article has to do with a special type of system called a *combined* system and the work that can be done by it. The combined system is made up of a control mass and an environment (Fig. 3-1).

If the control mass is at the restricted dead state, there is complete thermal and mechanical equilibrium throughout the combined system, and no work can be extracted from the combined system. When the control mass is at any other state than the dead state, a potential exists for extracting work from the combined system. Availability is a measure of this potential.

The points listed below are developed in the discussion to follow.

- As the control mass passes from a given state to the dead state there is a maximum amount of work that can be done by the combined system.
- The maximum work is extracted only when, as the control mass passes to the dead state, the combined system undergoes an internally reversible process. The maximum work is independent of the particular internally reversible process executed.
- The maximum work is called the *availability*.
- Availability, symbolized by A, is an extensive property calculated by

$$A = (E - U_o) + p_o(V - V_o) - T_o(S - S_o) \qquad (3\text{-}4\text{a})$$

where $E(= U + KE + PE)$, V, and S denote, respectively, the energy, volume, and entropy of the control mass at the given state and U_o, V_o, and S_o are the same properties when the control mass is at rest at the dead state.

Derivation. Figure 3-1 shows a combined system. Heat and work interactions can take place between the control mass and environment, but only

3-4 Fundamentals

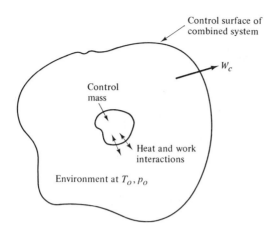

Figure 3-1 Combined system.

work interactions are permitted across the control surface of the combined system. And although the volumes of the control mass and environment can vary, the total volume remains constant: $\Delta(V + V^o) = 0$.

As the control mass passes from some given state to the dead state, the work derived from the combined system is determined from an energy equation as

$$W_C = -\Delta E_C$$

where the subscript C denotes the combined system. ΔE_C is the energy change of the combined system: the sum of the changes for control mass and environment

$$\Delta E_C = (U_o - E) + \Delta U^o$$

The energy change of the environment is given by Eq. (3-2a).

Collecting results, the work of the combined system is

$$W_C = -[(U_o - E) + (T_o \Delta S^o - p_o \Delta V^o)]$$

which, with $\Delta V^o = -\Delta V$ (assumption of constant total volume), becomes

$$W_C = (E - U_o) + p_o(V - V_o) - T_o \Delta S^o \qquad (a)$$

An entropy equation for the combined system gives

$$\Delta S_C = \sigma_C$$

ΔS_C is the sum of the entropy changes for the control mass and environment: $\Delta S_C = \Delta S + \Delta S^o$. So

$$(S_o - S) + \Delta S^o = \sigma_C \qquad (b)$$

The term σ_C accounts for the entropy produced due to irreversibilities within the combined system.

Combining Eqs. (a) and (b)

$$W_C = [(E - U_o) + p_o(V - V_o) - T_o(S - S_o)] - T_o\sigma_C \qquad (3\text{-}4\text{b})$$

Since $\sigma_C \geq 0$, it follows that

$$W_C \leq (E - U_o) + p_o(V - V_o) - T_o(S - S_o)$$

with the maximum work of the combined system being,

$$(W_C)_{\max} = (E - U_o) + p_o(V - V_o) - T_o(S - S_o) \qquad (3\text{-}4\text{c})$$

The maximum is obtained only when the process of the combined system is in every respect internally reversible. By definition, $A = (W_C)_{\max}$.

Availability cannot be less than zero. If the control mass is in any state other than the dead state, it is able to change its condition spontaneously toward the dead state; this tendency ceases when the dead state is reached. No work needs to be done on the combined system to effect such a spontaneous change. Accordingly, any change in state of the control mass to the dead state can be accomplished with at least zero work being derived from the combined system. It follows, then, that the *maximum* work extractable (availability) cannot be less than zero.

For the combined system to undergo an internally reversible process requires, in particular, that heat transfers of energy between the control mass and environment be made reversibly. This means that the temperature T of any part of the control mass undergoing a heat interaction with the environment is equal to (infinitesimally different than) that of the environment T_o, or that the heat interaction is effected through a reversible cycle operating between T and T_o. The work developed by the cycle contributes to the work output of the combined system. Since the working substance of the cycle is eventually restored to its initial state, the work developed is due solely to the heat transfers during the cycle and not from a change in state of the substance. Whether or not it would be economic to provide such a device is not important for the present development; the only issue is the exploitation of the combined system for work. The role of reversible cycles within this context is examined further below with examples.

Discussion. Availability is the maximum amount of work obtainable from the combined system as the control mass is brought into equilibrium with the environment. Its magnitude, which cannot be negative, depends on two states—the state of the control mass and that of the environment—and is a measure of the departure of the state of the control mass from that of the environment. Accordingly, availability has a positive value when the temperature T of the control mass is less than *or* greater than the temperature T_o. Similarly, it is positive when the pressure p of the control mass is less than *or* greater than p_o.

3-4 Fundamentals

This is brought out in the examples to follow. The first of the examples also illustrates that the value for availability is independent of the process selected to evaluate it and that the introduction of a reversible cycle is not essential.

Example 3-1. Consider a control mass at rest relative to the environment ($KE_t = PE_t = 0$) consisting of a simple compressible substance at $T_i < T_o$ and $p_i < p_o$, where T_o and p_o denote, respectively, the temperature and pressure of the environment. Evaluate the availability as the control mass is brought from this state to the dead state in each of the internally reversible processes described in parts (a) and (b).

(a) Figure E3-1a shows the states of the control mass. Process i–a is adiabatic and process a–o is at fixed volume.

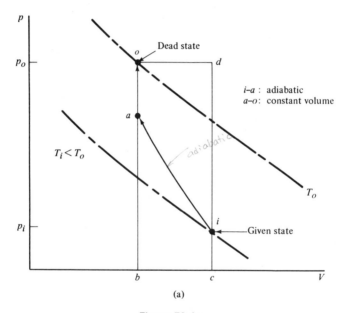

Figure E3-1a

Solution. Process i–a is an adiabatic and internally reversible process—that is, a constant entropy or *isentropic* process. Work must be done to compress the control mass

$$W = \int_i^a p\,dV < 0$$

The magnitude of this work is the enclosed area i–a–b–c–i. As the control mass is compressed, the environment expands and does work in amount $p_o(V_i - V_o)$. This corresponds to the rectangular area o–b–c–d–o. More work is done by the environment as it expands than is required to compress the control mass. The net amount, represented by the enclosed area o–d–i–a–o, is the amount available for transfer out of the combined system. This is expressed simply as

$$W_C = W - p_o \Delta V > 0 \tag{a}$$

where $\Delta V = V_o - V_i$. From an energy equation on the control mass $W = -(U_a - U_i)$; so

$$W_C = -(U_a - U_i) - p_o(V_o - V_i) > 0 \tag{b}$$

Process a–o is at constant volume. Neither the control mass nor the environment can do pdV-type work. However, a heat transfer of energy occurs and this can be exploited as follows.

The dead state can be reached by letting the control mass be heated at constant volume from state a by the environment. Work can be derived from the combined system for this process if the heat transfer is effected through a heat engine operating as shown in Fig. 3-2a.

The contribution of process a–o to the total work derived from the combined system is

$$W_C = \int_a^o \left(\frac{T_o}{T} - 1\right) \delta Q > 0$$

Inserting $\int_a^o \delta Q/T = S_o - S_i$ ($S_i = S_a$ for isentropic process i–a), along with $Q = U_o - U_a$, obtained from an energy equation for the control mass, results in

$$W_C = T_o(S_o - S_i) - (U_o - U_a) > 0 \tag{c}$$

Adding Eqs. (b) and (c) gives the total work done by the combined system,

$$W_C = (U_i - U_o) + p_o(V_i - V_o) - T_o(S_i - S_o) \tag{d}$$

which upon comparison with Eq. (3-4a) is seen to be the availability for a control mass at rest relative to the environment. In both parts of this calculation, the energy derived from the combined system as work originates in the energy of the environment.

(b) Figure E3-1b shows the states of the control mass. Process i–a is adiabatic, process a–o is isothermal to T_o.

3-4 Fundamentals

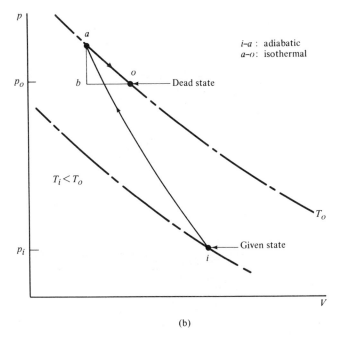

(b)

Figure E3-1b

Solution. For process i–a, the same reasoning as in part (a) gives

$$W_C = W - p_o \Delta V$$
$$= -(U_a - U_i) - p_o(V_a - V_i) \qquad \text{(e)}$$

Process a–o is an isothermal expansion to the dead state in which the control mass experiences both heat and work interactions. Heat transfer between the control mass and environment occurs at T_o and no heat engine need be introduced (were one assumed, no net work could be derived since the engine would operate between a zero temperature difference). The heat transfer experienced by the control mass for the internally reversible, isothermal process is $Q = T_o(S_o - S_i)$ ($S_i = S_a$, since i–a is isentropic).

Work can be derived from the combined system due to pdV-type work of the control mass and environment. Using the same reasoning as in part (a)

$$W_C = W - p_o(V_o - V_a)$$

(represented by the enclosed area a–o–b–a), where W is the work of the control mass which is determined from an energy equation as

$$W = Q - (U_o - U_a)$$

Collecting results

$$W_c = -(U_o - U_a) - p_o(V_o - V_a) + T_o(S_o - S_i) \tag{f}$$

Adding Eqs. (e) and (f) gives the same result as obtained in part (a), but without the need to introduce a heat engine.

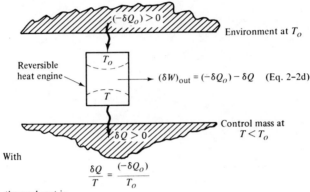

With

$$\frac{\delta Q}{T} = \frac{(-\delta Q_o)}{T_o}$$

the work out is,

$$(\delta W)_{out} = [\frac{T_o}{T} - 1] \, \delta Q$$

(a) Heat transfer from the environment to the control mass. $\delta Q > 0, T < T_o$

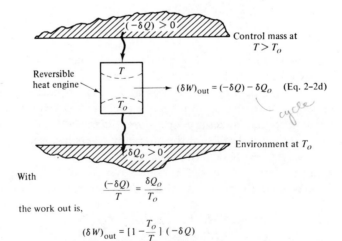

With

$$\frac{(-\delta Q)}{T} = \frac{\delta Q_o}{T_o}$$

the work out is,

$$(\delta W)_{out} = [1 - \frac{T_o}{T}] \, (-\delta Q)$$

$$= [\frac{T_o}{T} - 1] \, \delta Q$$

(b) Heat transfer from the control mass to the environment. $\delta Q < 0, T > T_o$

Figure 3-2 Reversible heat engines.

3-4 Fundamentals

evaluated as

$$\int_T^{T_o} \left(\frac{T_o}{T} - 1\right) \delta q$$

where δq, the heat transfer per unit of mass, is determined from an energy equation. It is left as an exercise to show that this procedure leads to the same value for the specific availability as calculated previously.

Availability Destruction. Unlike energy, availability is not conserved. It is destroyed by irreversibilities. A limiting case is when the availability is completely destroyed. This arises when the control mass changes to the dead state and no work is done by the combined system (as in a spontaneous change). The potential to deliver work the combined system had in its given condition is completely wasted.

More generally, there is some work done by the combined system and some destruction. This is brought out by combining Eqs. (3-4b) and (3-4c) to obtain

$$W_C = (W_C)_{\text{max}} - T_o \sigma_C$$

The effect of irreversibilities, embodied in the entropy production σ_C, is to keep the work below the maximum amount.

The destruction of availability is seen to be proportional to the creation of entropy due to irreversibilities. For easy reference, the irreversible destruction of availability is denoted by I

$$I = T_o \sigma$$

and is called the *irreversibility*.

The example below illustrates the availability destruction concept.

Example 3-4. As shown in Fig. E3-4, a control mass consists of two subsystems. One is N moles of an ideal gas at temperature $T_o + \Delta T$ and volume

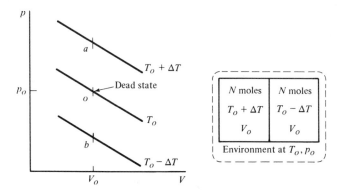

Figure E3-4

$V_o(= N\bar{R}T_o/p_o)$, state a on the property diagram, and the other is N moles of the same gas at temperature $T_o - \Delta T$ and volume V_o, state b on the diagram. Show that availability is destroyed if each subsystem attains the dead state at fixed volume while undergoing a heat interaction with the other.

Solution. The combined system of control mass and environment has a potential to do work. As suggested by the solutions to Examples 3-1 and 3-2, work can be developed as each subsystem separately comes into equilibrium with the environment. This potential is completely wasted in the spontaneous heat interaction under consideration.

Let the two subsystems have a spontaneous heat interaction with one another until a uniform temperature is achieved throughout the control mass. Assuming no other heat or work interactions, an energy equation for the control mass reduces to $\Delta U = 0$, from which the final temperature is found to be T_o. Thus, when the spontaneous process ceases, each subsystem is at the dead state, but no work is derived. The potential is completely wasted.

Summary. This section closes, as it begins, with a listing of important points related to the availability concept.

- Availability is the maximum work that can be extracted from the combined system of control mass and environment as the control mass passes from a given state to the dead state.
- At the dead state, both control mass and environment possess energy, but the availability is zero.
- For all states $A \geq 0$, where A denotes availability.
- Availability is an extensive property. Its value is fixed by the state of the control mass for any given values of p_o and T_o. The numerical value of availability depends explicitly on the choices for p_o and T_o.
- Availability is a measure of the departure of the state of control mass from that of the environment. The greater the difference between the temperature T at a given state and that of the environment T_o, the greater the value for availability. This applies equally when $T > T_o$ and $T < T_o$. Similarly, the greater the difference between pressure p at a given state and that of the environment p_o, the greater the availability.
- The kinetic and potential energies of the control mass (measured relative to the environment) contribute their full values to the availability magnitude, for in principle each is completely convertible to work as the control mass is brought to rest relative to the environment.
- Availability is destroyed by irreversibilities.

The property called availability developed in this section is based on the *restricted* dead state. At the restricted dead state the control mass is in *thermal*

3-5 Availability Equation for a Control Mass

and *mechanical* equilibrium with the environment. In recognition of this, it can be referred to as *thermomechanical* availability, to distinguish it from the more general concept developed in Chaps. 6 and 7 where the contents of the control mass are permitted to mix with the environment or enter into chemical reaction with environmental components and in so doing develop additional work.

3-5 Availability Equation for a Control Mass

Introduction. A control mass in a given state can attain new states through heat and/or work interactions with other systems. The other systems need not include the environment. Since the value of the availability at a new state will generally differ from that at the given state, it follows that availability can be altered by means of such interactions. The relationship between availability change and energy input is explored in this section first by considering special cases and then through a general formulation.

Heat Interaction. Let a control mass, initially at the dead state, be heated or cooled at constant volume in an interaction with some other system. The heat transfer experienced by the control mass is Q. Whether the mass is heated or cooled, the final state, f, has associated with it a positive value for availability because the temperature now differs from that of the environment. Thus, an increase in availability is realized by heating to increase the temperature from T_o to T_f, or by cooling to decrease the temperature from T_o to T_f. The availability associated with the final state equals the maximum work extractable from the combined system of control mass and environment as the mass is returned to the dead state (Sec. 3-4).

Consider first the case where the control mass undergoes an internally reversible process during the heat interaction with the other system. To evaluate the availability associated with the final state, f, any internally reversible process back to the dead state can be selected. A convenient choice is to let the control mass be cooled (case $T_f > T_o$) or heated (case $T_f < T_o$) by the environment to the dead state in an internally reversible manner at fixed volume.

Work can be derived in the process by means of a heat engine operating as shown in Fig. 3-2. The amount developed is

$$\int_f^o \left(\frac{T_o}{T} - 1\right) \delta \tilde{Q}$$

where $\tilde{Q}$ is the heat transfer experienced by the control mass in the interaction with the heat engine. Since both processes of the control mass, o–f and f–o, follow the same path but in reverse order

$$\int_f^o \frac{\delta \tilde{Q}}{T} = -\int_o^f \frac{\delta Q}{T}, \qquad \tilde{Q} = -Q$$

With these, the work developed can be expressed as

$$\int_o^f \left(1 - \frac{T_o}{T}\right) \delta Q$$

This expression gives the availability at the final state. It is natural to identify it as the *flow of availability associated with the heat transfer Q*.

If there is availability destruction within the control mass during the heat interaction, the flow of availability associated with the heat transfer is evaluated in the same way

$$\int_o^f \left(1 - \frac{T_o}{T_s}\right) \delta Q$$

where T_s is introduced to stress that this is the temperature on the control surface at the place where heat transfer occurs. But the availability value at the final state is necessarily less than when there is no destruction; that is, the change in availability from the initial to the final state, ΔA, satisfies

$$\Delta A < \int_o^f \left(1 - \frac{T_o}{T_s}\right) \delta Q$$

The availability change can also be shown as

$$\Delta A = \int_o^f \left(1 - \frac{T_o}{T_s}\right) \delta Q - I$$

where I accounts for the availability destruction within the control mass due to internal irreversibilities.

Work Interaction. Let a control mass, initially at the dead state, be compressed adiabatically in a work interaction with some other system. Retaining the usual sign convention, the work done *on* the control mass is $-W$. The availability associated with the control mass at the final state, f, is not identical to the work done on it in the interaction. Rather, the availability equals the maximum work extractable from the combined system of control mass and environment as the mass is returned to the dead state.

Consider first the case where the control mass undergoes an internally reversible process during the work interaction with the other system. To evaluate the availability associated with the final state, f, any internally reversible process back to the dead state can be selected. A convenient choice is an adiabatic process. The work developed is

$$\widetilde{W} - p_o(V_o - V_f)$$

where $\widetilde{W}$ is the work done by the mass as it returns to the dead state and $p_o(V_o - $

3-5 Availability Equation for a Control Mass

V_f) accounts for the work done in pushing aside the environment. Since $\widetilde{W} = -W$, the last expression may be rewritten as

$$-[W - p_o(V_f - V_o)]$$

This can be identified as the *flow of availability associated with the work W*.

If there is availability destruction within the control mass during the work interaction, the flow of availability is evaluated the same way, but the availability value at the final state is necessarily less than when there is no destruction; that is, the availability change from the initial to the final state, ΔA, satisfies

$$\Delta A < -[W - p_o(V_f - V_o)]$$

This can also be shown as

$$\Delta A = -[W - p_o(V_f - V_o)] - I$$

where I accounts for the availability destruction within the control mass due to internal irreversibilities.

Availability Equation. In the general case, a control mass may experience both heat and work interactions with other systems, not necessarily including the environment, and there is some availability destruction within it during such interactions. The availability equation, Eq. (3-5b), accounts for all these effects.

Let a control mass undergo heat and/or work interactions with other systems. The energy and entropy equations for a differential step of the process of the control mass are, respectively

$$dE = \delta Q - \delta W \quad \text{(a)}$$

$$dS = \frac{\delta Q}{T_s} + \delta \sigma \quad \text{(b)}$$

where T_s is the temperature on the control surface where the heat transfer occurs and $\delta\sigma$ accounts for entropy produced due to irreversibilities *within* the control mass.

Adding the term $p_o dV$ to both sides of Eq. (a), multiplying Eq. (b) by T_o, and subtracting the two resulting expressions gives

$$dE + p_o dV - T_o dS = \left(1 - \frac{T_o}{T_s}\right)\delta Q - (\delta W - p_o dV) - T_o \delta\sigma$$

Referring to Eq. (3-4a), the left side of the previous equation is seen to be the change in availability, dA. With this

$$dA = \left(1 - \frac{T_o}{T_s}\right)\delta Q - (\delta W - p_o dV) - T_o \delta\sigma \quad (3\text{-}5\text{a})$$

Integrating Eq. (3-5a) for a process from state 1 to 2

$$\Delta A = \underbrace{\int_1^2 \left(1 - \frac{T_o}{T_s}\right)\delta Q}_{(i)} - \underbrace{(W - p_o \Delta V)}_{(ii)} - \underbrace{I}_{(iii)} \qquad (3\text{-}5b)$$

where $\Delta A = \Delta E + p_o \Delta V - T_o \Delta S$. This is the availability equation for a control mass. Term (i) accounts for the flow of availability associated with heat transfer. Term (ii) is the availability flow with the work interaction. Term (iii) is the irreversibility $I(=T_o\sigma)$ which accounts for the destruction of availability due to irreversibilities within the control mass.

The availability equation may be written as a rate equation

$$\frac{dA}{dt} = \int_\alpha \left(1 - \frac{T_o}{T_s}\right) q_s d\alpha - (\dot{W} - p_o \dot{V}) - \dot{I} \qquad (3\text{-}5c)$$

where T_s is the temperature at that location of the control surface where the heat flux is q_s, and the integration is over the control surface. An important special case is

$$\frac{dA}{dt} = \sum_{j=1}^n \left(1 - \frac{T_o}{T_j}\right) \dot{Q}_j - (\dot{W} - p_o \dot{V}) - \dot{I} \qquad (3\text{-}5d)$$

where T_j is the temperature at that portion of the control surface where the heat transfer rate is $\dot{Q}_j$.

Discussion. It is easy to show that the principal results of the two special cases considered earlier in this section are readily obtained by reduction of Eq. (3-5b). This is left as an exercise.

The following should be noted about the flow of availability with heat transfer. If the heat transfer of energy takes place at a temperature above T_o, it is the system receiving energy that gains availability and the energy source that loses it (term (i) of Eq. (3-5b) is positive when $\delta Q > 0$ and $T_s > T_o$). Just the reverse is the case for a heat transfer of energy taking place at a temperature below T_o. That is, the system receiving energy loses availability and the energy source gains it (term (i) is negative when $\delta Q > 0$ and $T_s < T_o$).

The two examples to follow illustrate the use of the availability equation for a control mass.

Example 3-5. Determine the irreversibility for the case described in Example 2-1 and interpret the result. Let $T_1 > T_2 > T_o$.

Solution. Applying Eq. (3-5b) to the rod

$$0 = \left(1 - \frac{T_o}{T_1}\right)Q - \left(1 - \frac{T_o}{T_2}\right)Q - I$$

3-5 Availability Equation for a Control Mass

since no work is done and the rod is at steady-state. Solving for I and reducing

$$I = \left(1 - \frac{T_o}{T_1}\right)Q - \left(1 - \frac{T_o}{T_2}\right)Q$$
$$= T_o\left[\frac{Q}{T_1 T_2}(T_1 - T_2)\right] \quad (a)$$

The same result is obtained by combining the definition $I = T_o\sigma$ with the expression for the entropy production derived in Example 2-1.

The irreversibility in this case is due to heat transfer through a finite temperature difference. Eq. (a) shows the penalty exacted by the irreversibility: The term $[1 - (T_o/T_1)]Q$ is equivalent to the work that could be obtained from Q by an ideal heat engine operating between T_1 and T_o. Likewise, $[1 - (T_o/T_2)]Q$ is equivalent to the work that could be obtained from Q by an engine operating between T_2 and T_o. The difference between the two represents the potential irrevocably lost in the process; that is, represents the destruction of availability.

Example 3-6. Consider one lb of air contained in a closed, rigid, insulated tank at a temperature of 500°R and a pressure of 1 atm. Eighty-five Btu of paddlewheel work is done on the air. Assuming the ideal gas model with $c_v = 0.17$ Btu/lb°R, determine the irreversibility in Btu/lb and discuss the result. Let $T_o = 500°R$, $p_o = 1$ atm.

Solution. The final temperature of the air is determined from an energy equation

$$W/m = -\Delta u$$
$$= -c_v(T_2 - T_1)$$

Rearranging and inserting values

$$T_2 = T_1 - (W/mc_v) = 500 - (-85/(0.17))$$
$$= 1000°R$$

Next, the availability equation, Eq. (3-5b), reduces for the adiabatic, constant volume process to

$$I = -\Delta A - W$$

Evaluating ΔA via Eq. (3-4a) for the constant volume process

$$I/m = -(\Delta u - T_o\Delta s) - (W/m)$$

Introducing the energy equation, this reduces to

$$I/m = T_o \Delta s$$

which is of course just $I/m = T_o(\sigma/m)$.

Evaluating Δs, simplifying and inserting values

$$I/m = T_o c_v \ln(T_2/T_1)$$
$$= 500\,(0.17)\ln(2) = 58.9 \text{ Btu/lb}$$

The irreversibility in this case is due to fluid friction within the tank. As a result of the stirring action of the paddlewheel, the energy entering is stored as internal energy. However, the stored internal energy is not fully convertible to work, even by use of an ideal heat engine (see Problem 3-4). Energy is conserved (85 Btu of work is done and the internal energy increases by 85 Btu) but availability is destroyed.

3-6 Control Volume Availability Equation

The objective of this section is to extend the availability concept for application to control volumes. The presentation closely parallels the development of the control volume energy equation in Sec. 2-2.

Derivation. Figure 3-3 (Fig. 2-1 repeated) shows a control mass which at time t is the sum of the mass within the control volume and the mass δm within the shaded region adjacent to section i. The intensive properties of the mass δm are assumed to be those associated with section i. During time interval dt

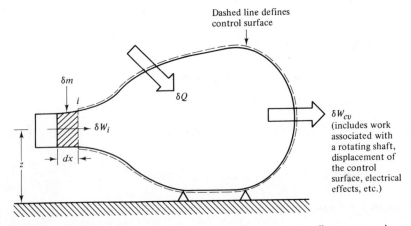

Figure 3-3 Control volume showing heat and work interactions as well as mass crossing the control surface.

3-6 Control Volume Availability Equation

the mass δm crosses i to join the matter initially within the control volume. At time $t + dt$ the boundary of the control mass corresponds exactly with that of the control volume.

The availability equation for the control mass is

$$A_{cm}(t + dt) - A_{cm}(t) = \left(1 - \frac{T_o}{T_s}\right)\delta Q - (\delta W - p_o dV) - \delta I \qquad (a)$$

As discussed in Sec. 2-2, the work term has two contributions. One of these lumps together all work interactions taking place at those portions of the control surface through which there is no mass flow. It is symbolized by δW_{cv}. The other is the *flow* work: the work, δW_i, required to push mass δm across section i. It is given by

$$\delta W_i = (p_i v_i)\delta m$$

where v_i is the specific volume and p_i the pressure at i.

The change in volume of the control mass is

$$dV = dV_{cv} - v_i \delta m \qquad (b)$$

where $-v_i \delta m$ is the volume change which occurs when δm crosses i and dV_{cv} accounts for volume change at portions of the control surface other than at i.

Accounting for the two work contributions separately and using Eq. (b), Eq. (a) becomes

$$A_{cm}(t + dt) - A_{cm}(t) = \left(1 - \frac{T_o}{T_s}\right)\delta Q - (\delta W_{cv} - p_o dV_{cv})$$
$$+ (p_i v_i - p_o v_i)\delta m - \delta I \qquad (c)$$

At time t the availability associated with the control mass is that of the control volume at time t plus that of the mass δm

$$A_{cm}(t) = A_{cv}(t) + a_i \delta m$$

where a_i is the specific availability accompanying the mass entering at i. At time $t + dt$

$$A_{cm}(t + dt) = A_{cv}(t + dt)$$

Inserting these into Eq. (c) and grouping terms

$$dA_{cv} = \left(1 - \frac{T_o}{T_s}\right)\delta Q - (\delta W_{cv} - p_o dV_{cv}) + [a_i + (p_i - p_o)v_i]\delta m - \delta I \qquad (d)$$

where $dA_{cv} = A_{cv}(t + dt) - A_{cv}(t)$.

Flow Availability. The availability of the control volume can be altered by interactions at the flow port i. The term which accounts for this is $[a_i + (p_i - p_o)v_i]\delta m$, where a_i is the specific availability associated with the flowing mass and $(p_i - p_o)v_i$ is the contribution of the work interaction at i.

It is convenient to have a distinct symbol and designation for this sum. The (thermomechanical) *flow availability* at port i, a_{fi}, is defined as

$$a_{fi} = a_i + (p_i - p_o)v_i \qquad (e)$$

Comparison of the current development with that of Sec. 2-2 shows that the flow availability evolves here in the same way as does enthalpy in the control volume energy equation derivation. And each has a similar interpretation: It is a sum consisting of a term associated with the flowing mass (specific internal energy for enthalpy, specific availability for flow availability) and a contribution associated with the work interaction at the flow port.

The flow availability can be placed in a form convenient for calculation by using Eq. (3-4a) to write

$$a_i = (e_i - u_o) + p_o(v_i - v_o) - T_o(s_i - s_o)$$

Then, upon eliminating e_i with

$$e_i = u_i + \frac{\mathcal{V}_i^2}{2} + gz_i \qquad (2\text{-}2a)$$

and inserting the result into Eq. (e)

$$a_{fi} = \left[(u_i - u_o) + p_o(v_i - v_o) - T_o(s_i - s_o) + \frac{\mathcal{V}_i^2}{2} + gz_i \right] + (p_i - p_o)v_i$$

Finally, with $h = u + pv$

$$a_{fi} = (h_i - h_o) - T_o(s_i - s_o) + \frac{\mathcal{V}_i^2}{2} + gz_i \qquad (3\text{-}6a)$$

Though the specific availability a_i cannot be negative, the flow availability can take on negative values if the pressure at i is less than the pressure of the environment, $p_i < p_o$. In the literature, the flow availability is often referred to as *exergy*.

Rate Equation. On a time rate basis, and including the possibility of heat transfer through several portions of the control surface, Eq. (d) becomes

$$\frac{dA_{cv}}{dt} = \int_a \left(1 - \frac{T_o}{T_s}\right) q_s d\mathcal{A} - \left(\dot{W}_{cv} - p_o \frac{dV_{cv}}{dt}\right) + \dot{m}_i a_{fi} - \dot{I}$$

where T_s is the temperature at that location of the control surface where the heat flux is q_s and a_{fi} is the flow availability.

This reasoning can be extended to a control volume with matter flowing in and out at several locations to obtain

$$\frac{dA_{cv}}{dt} = \int_a \left(1 - \frac{T_o}{T_s}\right) q_s d\mathbf{Q} - \left(\dot{W}_{cv} - p_o \frac{dV_{cv}}{dt}\right) + \sum_i \dot{m}_i a_{fi} - \sum_e \dot{m}_e a_{fe} - \dot{I} \quad (3\text{-}6b)$$

A special case is

$$\frac{dA_{cv}}{dt} = \sum_{j=1}^{n} \left(1 - \frac{T_o}{T_j}\right) \dot{Q}_j - \left(\dot{W}_{cv} - p_o \frac{dV_{cv}}{dt}\right) + \sum_i \dot{m}_i a_{fi} - \sum_e \dot{m}_e a_{fe} - \dot{I} \quad (3\text{-}6c)$$

where T_j is the temperature at that portion of the control surface where the heat transfer rate is $\dot{Q}_j$.

Steady-State Forms. For a control volume at steady-state, Eqs. (3-6b and 3-6c) reduce, respectively, to

$$0 = \int_a \left(1 - \frac{T_o}{T_s}\right) q_s d\mathbf{Q} - \dot{W}_{cv} + \sum_i \dot{m}_i a_{fi} - \sum_e \dot{m}_e a_{fe} - \dot{I}$$

$$0 = \sum_{j=1}^{n} \left(1 - \frac{T_o}{T_j}\right) \dot{Q}_j - \dot{W}_{cv} + \sum_i \dot{m}_i a_{fi} - \sum_e \dot{m}_e a_{fe} - \dot{I} \quad (3\text{-}6d)$$

since at steady-state $dA_{cv}/dt = dV_{cv}/dt = 0$.

If there is only a single flow inlet and a single flow exit

$$0 = \int_a \left(1 - \frac{T_o}{T_s}\right) q_s d\mathbf{Q} - \dot{W}_{cv} + \dot{m}_f (a_{fi} - a_{fe}) - \dot{I}$$

$$0 = \sum_{j=1}^{n} \left(1 - \frac{T_o}{T_j}\right) \dot{Q}_j - \dot{W}_{cv} + \dot{m}_f (a_{fi} - a_{fe}) - \dot{I} \quad (3\text{-}6e)$$

where $\dot{m}_f$ is the mass flow rate.

Unlike the steady-state form of the control volume energy equation, Eq. (2-2i), which indicates that the rate energy enters the control volume equals the rate energy exits, Eqs. (3-6d) and (3-6e) show that the rate availability enters must *exceed* the rate it exits. The term $\dot{I}$ accounts for destruction of availability within the control volume due to internal irreversibilities.

3-7 Control Volume Applications

The objective of this section is to illustrate the use of the control volume availability equation while bringing out several important aspects of availability analysis.

For a careful evaluation of the performance of devices, both availability losses and irreversibilities should be considered. The availability equation provides a systematic and comprehensive approach for such evaluations. These ideas are emphasized in several places in this book and are brought out in the example to follow.

Example 3-7. A gas turbine power plant is shown in Fig. E3-7. It operates at steady-state with air as the working fluid. Property data are given in the table below. Using Appendix Table B-6A to evaluate properties, analyze the power plant on both an energy and an availability basis. Assume the turbine and compressor are adiabatic. Neglect kinetic and potential energy changes. Let $T_o = 80°F$, $p_o = 14.5 \text{ lbf/in.}^2$

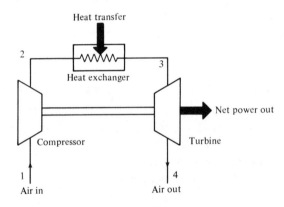

Figure E3-7

State	T (°F)	p (lbf/in.²)	h (Btu/lb)	a_f (Btu/lb)
1	80	14.5	129	0
2	514	87	235	95
3	1540	87	505	263
4	917	14.5	337	83

Solution. Using the tabulated temperatures, the enthalpy values reported in the table are found from Appendix Table B-6A. The flow availabilities are calculated from Eq. (3-6a)

$$a_f = (h - h_o) - T_o(s - s_o)$$
$$= (h - h_o) - T_o[(s' - s'_o) - R \ln (p/p_o)]$$

where s' is obtained from Appendix Table B-6A.

The tabulated enthalpy and flow availability values are employed to obtain

3-7 Control Volume Applications

	Energy (Btu/lb)	Availability (Btu/lb)
Increase through Heat Exchanger	270[a]	168[b]
Outputs:		
Net work	62[c]	62
Air out at 4	208[d]	83[e]
Irreversibility:		
Compressor	—	11[f]
Turbine	—	12
Total	270	168

[a] Value calculated from $(h_3 - h_2)$
[b] Value calculated from $(a_{f3} - a_{f2})$
[c] Value calculated from $(h_3 - h_4) - (h_2 - h_1)$
[d] Value calculated from $(h_4 - h_1)$
[e] Value calculated from $(a_{f4} - a_{f1})$
[f] $\dot{I}/\dot{m}_f = -(\dot{W}/\dot{m}_f) + a_{f1} - a_{f2}$ (from Eq. (3-6e))

Comparison of values shows that the energy analysis gives a different picture of performance than the availability analysis. The energy analysis is silent about the effect of irreversibilities within the compressor and turbine. It draws attention mainly to the large "loss" associated with the effluent, the exiting air. The availability analysis shows that the relative significance of this loss is less, and also stresses that internal irreversibilities exact a penalty as well. To improve overall performance, both losses and irreversibilities must be considered.

As shown by the previous example, the availability equation enables a careful distinction to be drawn between losses and destructions of availability, and permits a correct evaluation of their relative significance. Accordingly, it can be used to identify and rank order those sites where important losses and destructions occur. The list so generated serves as a guide for deciding which areas offer the greatest opportunities for improvement through application of practical engineering measures.

It is frequently instructive to express availability losses and destructions in terms of key engineering parameters, and in particular *dimensionless* parameters. The following illustration develops this idea within a simple context.

Illustration. Consider the insulated pipe shown in Fig. 3-4 through which an incompressible fluid is flowing steadily. For the element dL shown in the figure expressions in terms of key engineering parameters are developed for the loss of availability due to heat transfer and the irreversibility. The results are Eqs. (3-7a) and (3-7b). The presentation concludes with a brief discussion of these equations.

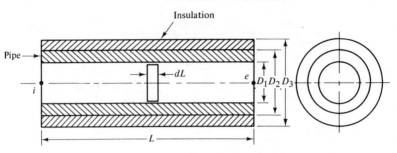

Figure 3-4 Insulated pipe.

For the element dL, the loss of availability due to heat transfer is

$$\delta \dot{A}_{\text{loss}} = \left(1 - \frac{T_o}{T_f}\right)\delta \dot{Q}$$

where T_f is the local fluid temperature and $\delta\dot{Q}$ is a *positive* number accounting for the time rate of heat transfer *from* the pipe. Drawing upon heat transfer fundamentals, $\delta\dot{Q}$ can be evaluated as[2]

$$\delta\dot{Q} = \frac{\pi dL(T_f - T_\infty)}{\left[\frac{1}{2k_p}\ln\left(\frac{D_2}{D_1}\right) + \frac{1}{2k_i}\ln\left(\frac{D_3}{D_2}\right) + \frac{1}{h_\infty D_3}\right]} \qquad (a)$$

where k_p and k_i denote, respectively, the *thermal conductivities* of the pipe and insulation. T_∞ is the temperature of the surroundings. The letter D denotes diameters.

In Eq. (a), h_∞ is the *convective heat transfer coefficient* of the surrounding media. It can be eliminated in terms of the dimensionless *Nusselt* number, Nu,

$$\text{Nu} = \frac{h_\infty D_3}{k_\infty}$$

where k_∞ is the thermal conductivity. For *forced* convection heat transfer, the Nusselt number depends on the *Reynolds* and *Prandtl* numbers. In *free* convection, the *Grashof* number replaces the Reynolds number. For otherwise similar cases, the Nusselt number is smaller in value in free convection than in forced convection.[3]

---

[2] Frank Krieth, *Principles of Heat Transfer*, 3rd ed., Intext Educational Publishers, New York, 1973, Sec. 2-2. Radiation is neglected.

[3] Frank Krieth, *Principles of Heat Transfer*, 3rd ed., Intext Educational Publishers, New York, 1973, Secs. 7-3 and 9-2.

3-7 Control Volume Applications

Collecting results

$$\delta \dot{A}_{\text{loss}} = \frac{\pi dL\left(1 - \frac{T_o}{T_f}\right)(T_f - T_\infty)}{\left[\frac{1}{2k_p}\ln\left(\frac{D_2}{D_1}\right) + \frac{1}{2k_i}\ln\left(\frac{D_3}{D_2}\right) + \frac{1}{k_\infty \text{Nu}}\right]} \tag{3-7a}$$

To evaluate the irreversibility associated with flow through the pipe, apply Eq. (3-6e) to the element dL

$$\delta \dot{I} = \left(1 - \frac{T_o}{T_f}\right)(-\delta \dot{Q}) - \dot{m}_f da_f$$

where $da_f = dh - T_o ds$. Eliminating $\delta \dot{Q}$ by use of an energy equation $((-\delta \dot{Q}) = \dot{m}_f dh)$, and using $dh = T_f ds + v dp$ (Eq. 2-4d), this reduces to

$$\delta \dot{I} = -\frac{\dot{m}_f T_o}{T_f}\frac{dp}{\rho} \tag{b}$$

where ρ is the fluid density.

The pressure drop can be eliminated in favor of the *friction factor f*.

$$(-dp/\rho) = \mathcal{V}^2 f dL/2D.$$

The friction factor depends on the Reynolds number for tube flow and the ratio $\mathscr{e}/D_1$, where $\mathscr{e}$ is a measure of pipe roughness. f increases with $\mathscr{e}$ for fixed Reynolds number.[4] After reduction

$$\delta \dot{I} = \frac{T_o}{T_f}\left[\frac{8(\dot{m}_f)^3}{\pi^2 \rho^2 D_1^5}\right] f dL \tag{3-7b}$$

Study of Eqs. (3-7a) and (3-7b) leads to familiar conclusions about performance. If the numerator of Eq. (3-7a) is regarded as fixed, the loss can be kept low by resorting to pipe and insulation materials having low thermal conductivities, a relatively thick walled pipe (large D_2/D_1), and relatively thick insulation (large D_3/D_2). Forced convection cooling on the outer surface should be avoided. For a specified flow rate, Eq. (3-7b) shows that the irreversibility can be kept low by using a relatively smooth walled pipe having a large inner diameter.

This illustration is considered further in Sec. 9-4 from the perspective of *thermoeconomics*.

The Irreversibility. There are a number of possible avenues open when the objective is to calculate the irreversibility. One of these is the direct use of the

---

[4]Frank Krieth, *Principles of Heat Transfer*, 3rd ed., Intext Educational Publishers, New York, 1973, Sec. 8-2.

availability equation. For example, in the case of a control volume at steady-state with one flow inlet and one flow exit, Eq. (3-6e) gives

$$\dot{I} = \int_\alpha \left(1 - \frac{T_o}{T_s}\right) q_s d\alpha - \dot{W}_{cv} + \dot{m}_f(a_{fi} - a_{fe}) \tag{c}$$

Another avenue is to determine the entropy production $\dot{\sigma}$ by means of an entropy equation and write $\dot{I} = T_o \dot{\sigma}$. In many cases this is simpler to implement than the direct use of the availability equation. That the two approaches are equivalent is easily seen by subtracting the energy equation

$$0 = \dot{Q} - \dot{W}_{cv} + \dot{m}_f\left[\left(h + \frac{\mathcal{V}^2}{2} + gz\right)_i - \left(h + \frac{\mathcal{V}^2}{2} + gz\right)_e\right]$$

from Eq. (c) and reducing to obtain[5]

$$\dot{I} = -T_o\left[\int_\alpha \frac{q_s}{T_s} d\alpha + \dot{m}_f(s_i - s_e)\right]$$
$$= T_o \dot{\sigma}$$

The last equation emphasizes that when the irreversibility is the objective the only property of the environment that *must* be specified is temperature T_o.

There is another approach to calculating the irreversibility which appears occasionally in the literature.[6] This is elicited upon noting that the quantity $(h_o - T_o s_o)$ cancels from the flow availability difference

$$a_{fi} - a_{fe} = \left[(h - h_o) - T_o(s - s_o) + \frac{\mathcal{V}^2}{2} + gz\right]_i$$
$$- \left[(h - h_o) - T_o(s - s_o) + \frac{\mathcal{V}^2}{2} + gz\right]_e$$
$$= \left[h - T_o s + \frac{\mathcal{V}^2}{2} + gz\right]_i - \left[h - T_o s + \frac{\mathcal{V}^2}{2} + gz\right]_e$$

and so it is possible to rewrite Eq. (c) as

$$\dot{I} = \int_\alpha \left(1 - \frac{T_o}{T_s}\right) q_s d\alpha - \dot{W}_{cv} + \dot{m}_f\left[\left(b + \frac{\mathcal{V}^2}{2} + gz\right)_i - \left(b + \frac{\mathcal{V}^2}{2} + gz\right)_e\right]$$

where b is defined as $h - T_o s$. Notice though that the term $(b + \mathcal{V}^2/2 + gz)$

---

[5] Reversing the order, Eq. (c) can be developed by multiplying the entropy equation by T_o, subtracting the result from an energy equation, and introducing $(h_o - T_o s_o)$ where required. This is similar to the procedure used in Sec. 3-5 to establish the control mass form of the availability equation.

[6] See for example Eq. (155) of Ref. [3].

3-7 Control Volume Applications

does not represent flow availability. Accordingly, caution should be exercised in using it for any other purpose than evaluating the irreversibility or a *difference* in flow availability.

In the remainder of this section a number of examples are presented in which the irreversibility is calculated by application of the availability equation. It is left as an exercise to show that the calculations can also be performed by direct use of the entropy production concept.

Example 3-8. Consider the steady throttling of superheated water vapor at $p_1 = 500$ lbf/in.² and $T_1 = 500°F$ to a pressure $p_2 = 80$ lbf/in.². Determine the specific flow availability at inlet and exit, and the irreversibility per unit mass flowing. Neglect heat losses and kinetic and potential energy effects. Let $p_o = 1$ atm, $T_o = 537°R$. The situation is shown in Fig. E3-8.

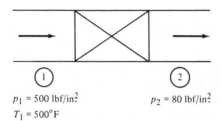

$p_1 = 500$ lbf/in.²
$T_1 = 500°F$

$p_2 = 80$ lbf/in.²

Figure E3-8

Solution. Reducing the steady-state mass and energy equations results in

$$h_2 = h_1$$

The state at the inlet is fixed by p_1, T_1 and at the exit by p_2, h_2. From Appendix Table B-2C, $h_1 = h_2 = 1231.5$ Btu/lb, $s_1 = 1.4923$ Btu/lb°R, $s_2 = 1.679$. As an approximation, properties at the dead state are evaluated at the saturated liquid state corresponding to T_o. From Appendix Table B-2A, $h_o = 45.09$, $s_o = 0.0859$.

With Eq. (3-6a)

$$a_{f1} = (h_1 - h_o) - T_o(s_1 - s_o)$$
$$= (1231.5 - 45.09) - 537(1.4923 - 0.0859)$$
$$= 431.2 \text{ Btu/lb}$$

Similarly,

$$a_{f2} = 330.9 \text{ Btu/lb}$$

Reducing Eq. (3-6e)

$$\frac{\dot{i}}{\dot{m}_f} = a_{f1} - a_{f2}$$
$$= 100.3 \text{ Btu/lb}$$

Energy is conserved in the throttling process, but availability is destroyed. The source of destruction is the uncontrolled expansion which occurs.

Example 3-9. Consider the steady mixing of a stream of liquid water at $T_1 = 200°F$ with a stream of liquid water at $T_2 = 60°F$. If the streams have equal mass flow rates, determine the temperature after mixing and the irreversibility in Btu/lb. Neglect heat losses and kinetic and potential energy effects. Let $T_o = 510°R$, $p_o = 1$ atm. See Fig. E3-9.

Figure E3-9

Solution. Reducing the steady forms of the mass and energy equations results in $h_3 = (h_1 + h_2)/2$.

Evaluating properties at the saturated liquid state for each specified temperature from Appendix Table B-2A, $h_1 = 168.07$ Btu/lb, $s_1 = 0.294$ Btu/lb°R, $h_2 = 28.08$, $s_2 = 0.0555$, $h_o = 18.06$, $s_o = 0.0361$.

With h_1 and h_2 known, the energy equation gives, $h_3 = 98.1$. Then, from Appendix Table B-2A, $T_3 = 130°F$, $s_3 = 0.1817$.

From Eq. (3-6a)
$$a_{f1} = (h_1 - h_o) - T_o(s_1 - s_o)$$
$$= 18.48 \text{ Btu/lb}$$

Similarly
$$a_{f2} = 0.13 \text{ Btu/lb} \quad \text{and} \quad a_{f3} = 5.76 \text{ Btu/lb}$$

Reducing Eq. (3-6d)
$$\frac{\dot{i}}{\dot{m}_3} = (a_{f1} + a_{f2})/2 - a_{f3}$$
$$= 3.55 \text{ Btu/lb}$$

Energy is conserved in the process, but availability is destroyed. Availability is destroyed when streams having different temperatures are mixed.

Example 3-10. Determine an expression for the irreversibility per unit mass of stream B for the counterflow heat exchanger pictured. Assume steady-state operation and the ideal gas model with constant specific heats. Further, neglect kinetic and potential energy changes, the pressure drop in each stream, and heat losses. See Fig. E3-10.

3-7 Control Volume Applications

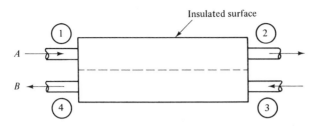

Figure E3-10

Solution. Reducing the steady-state form of the availability rate equation

$$\dot{I} = \dot{m}_A(a_{f1} - a_{f2}) + \dot{m}_B(a_{f3} - a_{f4})$$

$$\frac{\dot{I}}{\dot{m}_B} = \left[\left(\frac{\dot{m}_A}{\dot{m}_B}\right)(h_1 - h_2) + (h_3 - h_4)\right] - T_o\left[\left(\frac{\dot{m}_A}{\dot{m}_B}\right)(s_1 - s_2) + (s_3 - s_4)\right]$$

The first term on the right vanishes by application of an energy rate equation. Thus

$$\frac{\dot{I}}{\dot{m}_B} = T_o\left[\frac{\dot{m}_A}{\dot{m}_B}(s_2 - s_1) + (s_4 - s_3)\right]$$

The quantity in brackets is the entropy production per unit mass of stream B. Evaluating terms for the same fluid on each side

$$\frac{\dot{I}}{\dot{m}_B} = T_o c_p \left[\left(\frac{\dot{m}_A}{\dot{m}_B}\right) \ln\left(\frac{T_2}{T_1}\right) + \ln\left(\frac{T_4}{T_3}\right)\right]$$

$$= T_o c_p \ln\left[\left(\frac{T_2}{T_1}\right)^\alpha \left(\frac{T_4}{T_3}\right)\right]$$

where $\alpha = \dot{m}_A/\dot{m}_B$.

Energy is conserved, but availability is destroyed. The source of irreversibility here is heat transfer through a finite temperature difference. For further discussion of gas-to-gas counterflow heat exchangers see Problem 3-24.

Example 3-11. An air turbine has the following steady-state operating conditions. At the inlet the pressure is 75 lbf/in^2., the temperature is 800°R and the fluid velocity is 400 ft/s; at the exit, the conditions are 15 lbf/in^2., 600°R, and 100 ft/s. Assuming the ideal gas model with $c_p = 0.24$ Btu/lb°R, determine per unit mass passing through the turbine the work output and the irreversibility. Neglect heat transfer and potential energy. Let $T_o = 500°R$, $p_o = 15$ lbf/in.^{2}

Solution. Combining the steady-state forms of the mass and energy equations results in

$$\frac{\dot{W}_{cv}}{\dot{m}_f} = h_1 - h_2 + \frac{\mathcal{V}_1^2 - \mathcal{V}_2^2}{2}$$

$$= 0.24(800 - 600) + \frac{(400)^2 - (100)^2}{(2)(32.2)(778)}$$

$$= 48 + 3 = 51 \text{ Btu/lb}$$

Reducing Eq. (3-6e)

$$\dot{I}/\dot{m}_f = -\dot{W}_{cv}/\dot{m}_f + (a_{f1} - a_{f2})$$

where

$$a_{f1} - a_{f2} = (h_1 - h_2) - T_o(s_1 - s_2) + \frac{\mathcal{V}_1^2 - \mathcal{V}_2^2}{2}$$

$$= 48 - 500\left[0.24 \ln\left(\frac{800}{600}\right) - \frac{1.986}{29} \ln\left(\frac{75}{15}\right)\right] + 3$$

$$= 48 + 20.6 + 3$$

$$= 71.6 \text{ Btu/lb}$$

In this calculation $\bar{R} = 1.986$ Btu/(lbmol)(°R) (Appendix Table A-2) and the molecular weight of air is taken to be 29. Collecting results

$$\dot{I}/\dot{m}_f = -51 + 71.6$$

$$= 20.6 \text{ Btu/lb}$$

Discussion. For the adiabatic turbine, flow availability decreases from the inlet to exit because power is delivered and availability is destroyed due to internal irreversibilities. A parameter that gauges how effectively the turbine converts the flow availability decrease into work is

$$\epsilon = \frac{\dot{W}/\dot{m}_f}{a_{f1} - a_{f2}}$$

This is known as the turbine *effectiveness* and is a type of *second law efficiency* (see Chap. 4). Inserting values for the case under consideration

$$\epsilon = \frac{51}{71.6} = 0.71$$

The turbine effectiveness differs from the "isentropic" turbine efficiency η_t which compares the actual work developed to the work that would be developed in an isentropic expansion between the inlet state and the turbine exhaust pressure. The two efficiencies ϵ and η_t are discussed and compared in Ref. [3].

3-7 Control Volume Applications

Whether an unavoidable heat transfer is counted as a loss or a contributor to the irreversibility depends on the way the control surface is defined. In the example to follow, the control surface is located so that the unavoidable heat transfer is treated as a loss. However, by selecting the control surface to take in a portion of the immediate surroundings so that the heat transfer occurs at temperature T_o, it contributes to the irreversibility. The case of Example 3-12 is reconsidered in Problem 3-15; there the heat transfer contributes to the irreversibility of a suitably enlarged control volume.

Example 3-12. An air turbine has the following steady-state operating conditions. At the inlet the pressure is 75 lbf/in.^{2}, the temperature is 800°R and the fluid velocity is 400 ft/s; at the exit, the conditions are 15 lbf/in.^{2}, 600°R and 100 ft/s. There is a heat transfer of energy from the turbine in amount 10 Btu/lb across a portion of the control surface at $T_s = 700°R$. Assuming the ideal gas model with $c_p = 0.24$ Btu/lb°R, determine per unit mass passing through the turbine the work output and the irreversibility. Neglect potential energy. Let $T_o = 500°R$, $p_o = 15$ lbf/in.^{2}

Solution. Combining the steady-state forms of the mass and energy equation

$$\frac{\dot{W}_{cv}}{\dot{m}_f} = \frac{\dot{Q}}{\dot{m}_f} + (h_1 - h_2) + \left(\frac{\mathcal{V}_1^2 - \mathcal{V}_2^2}{2}\right)$$

$$= -10 + 48 + 3 = 41 \text{ Btu/lb}$$

Reducing Eq. (3-6e)

$$\frac{\dot{I}}{\dot{m}_f} = \left(1 - \frac{T_o}{T_s}\right)\left(\frac{\dot{Q}}{\dot{m}_f}\right) - \left(\frac{\dot{W}_{cv}}{\dot{m}_f}\right) + (a_{f1} - a_{f2})$$

$$= \left(1 - \frac{500}{700}\right)(-10) - 41 + 71.6$$

$$= -2.86 - 41 + 71.6$$

$$= 27.74 \text{ Btu/lb}$$

There is a flow of availability out associated with the heat transfer equal to 2.86 Btu/lb. The extent to which the potential of this "waste" heat is not utilized in the turbine's surroundings contributes to the irreversibility of the surroundings and not to that of the turbine. Nonetheless, in calculating the turbine *effectiveness* it can be regarded as a "loss" chargeable to the turbine performance: The flow availability decreases from inlet to exit because power is delivered, there is a transfer of availability out with heat transfer, and availability is destroyed due to internal irreversibilities. The turbine effectiveness evaluates how effectively the turbine converts the flow availability decrease to work; that is

$$\epsilon = \frac{\dot{W}/\dot{m}_f}{a_{f1} - a_{f2}}$$

$$= \frac{41}{71.6} = 0.57$$

3-8 Closure

The term *thermomechanical availability* is used to distinguish the availability concept introduced in this chapter from the more general concept developed in Chaps. 6 and 7. Though somewhat limited in scope, the thermomechanical availability is nonetheless useful for engineering analysis. This is brought out by the examples and illustrations considered within the chapter and by the end of chapter problems. The thermomechanical availability also suffices for an introduction to *second law efficiencies* and their use. This is the subject of the next chapter.

Selected References

1. BRUGES, E. A., *Available Energy and the Second Law Analysis*, Butterworths Scientific Publications, London, 1959.
2. HAYWOOD, R. W., "A Critical Review of the Theorems of Thermodynamic Availability, with Concise Formulations," *J. Mech. Engr. Sci.*, **16**, *3*, 1974, 160–173 and *4*, 1974, 258–267.
3. KEENAN, J. H., *Thermodynamics*, John Wiley & Sons, New York, 1941.
4. ———, "Availability and Irreversibility in Thermodynamics," *Brit. J. Appl. Phys.*, **2**, July 1951, 183–192.
5. OBERT, E. F., and R. A. GAGGIOLI, *Thermodynamics*, McGraw-Hill, New York 1963.

Problems

Unless otherwise noted, in the following problems take $T_o = 60°F$, $p_o = 14.7$ lbf/in.^{2}.

3-1 Repeat Example 3-1 using the following sequence of processes.
   i–x: isentropic compression to $T_x < T_o$
   x–y: constant volume to T_o
   y–o: isothermal compression to dead state

3-2 Consider a control mass at rest relative to the environment ($KE_i = PE_i = 0$) consisting of a simple compressible substance at $T_i > T_o$, $p_i < p_o$. Using the approach of Examples 3-1 and 3-2, evaluate the availability with each of the following internally reversible processes,
   (a) isentropic expansion to T_o, followed by an isothermal compression to the dead state.

(b) isentropic compression to volume equal to V_o, followed by constant volume cooling to the dead state.

3-3 Complete Example 3-3 as indicated.

3-4 Referring to Example 3-6, evaluate the availability at the final state. Show that this value equals the maximum work obtainable from a heat transfer of energy between the air and the environment via a power cycle as the air cools at constant volume to its initial state, the dead state.

3-5 Showing all important details, develop Eqs. (a), (3-7a), (b), and (3-7b) of Sec. 3-7 (see pp. 68–69).

3-6 Referring to Problem 2-14, calculate the net rate that availability exits with heat transfer and enters with work. Also, calculate the irreversibility in Btu/s. Let $T_o = 500°R$.

3-7 Determine the specific availability in Btu/lb for water at these states
(a) saturated liquid at 32°F
(b) saturated vapor at 212°F

3-8 A tank contains one lb of air at 58.8 lbf/in.^{2}, 200°F. Assuming the ideal gas model, determine the specific availability in Btu/lb.

3-9 A flywheel with moment of inertia 160 lb · ft^2 rotates at 3000 rpm. As the wheel is braked to rest, the rotational kinetic energy is converted by friction to internal energy of the brake lining. Assume there are no stray heat transfers with the surroundings. Take the brake lining as equivalent to 5 lb of a liquid with specific heat, $c = 1$ Btu/lb°R.
(a) Determine the final temperature of the brake lining if its initial temperature is T_o.
(b) Determine the maximum theoretical speed in rpm that could be attained by the flywheel by use of the internal energy stored in the brake lining after braking.

3-10 Assuming the ideal gas model, show that the availability of air stored in a closed tank of volume V at temperature T_o and pressure p is

$$A = p_o V \left[1 - \left(\frac{p}{p_o}\right) + \left(\frac{p}{p_o}\right) \ln \left(\frac{p}{p_o}\right) \right]$$

Find V in ft^3 if $A = 3412$ Btu (1 kWh), $p = 1000$ lbf/in.^{2}.

3-11 One lb of H_2O in a piston-cylinder device is initially at 160 lbf/in.^{2}, 400°F. A change in state occurs so that finally the pressure is 160 lbf/in.^{2} and $V_2 = 1.54\, V_1$.
(a) Determine the change in availability for this change in state.
(b) If the change in state occurs adiabatically by stirring the fluid, determine the irreversibility of the unit mass.
(c) If the change in state is brought about by heating the mass while maintaining its pressure constant, show that the value of the irreversibility is zero.

3-12 Consider a circular rod with constant thermal conductivity k whose ends are fixed at temperatures T_1, T_2 ($< T_1$) as shown in Fig. P3-12. The rate of heat transfer with the surroundings from its outer surface *per unit of length* is $\dot{q}'$. The rate $\dot{q}'$ may vary axially.
(a) Apply an energy equation to an elemental length dx of the rod and use Fourier's conduction equation

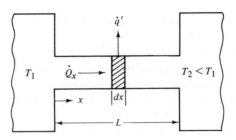

Figure P3-12

$$\dot{Q}_x = -kA\frac{dT}{dx} \qquad (a)$$

where A is the rod cross-sectional area, $\dot{Q}_x$ is the axial heat transfer rate through the rod and $T(x)$ is the temperature within the rod, to obtain

$$kA\frac{d^2T}{dx^2} = \dot{q}'$$

(b) Apply an entropy equation to the element to obtain the following expression for the rate of entropy production *per unit of length* $\dot{\sigma}'$

$$\dot{\sigma}' = \dot{Q}_x \frac{d}{dx}\left(\frac{1}{T}\right)$$

(c) Since the irreversibility rate *per unit of length* is $i' = T_o \dot{\sigma}'$

$$i' = T_o \dot{Q}_x \frac{d}{dx}\left(\frac{1}{T}\right)$$

Show that by use of Eq. (a), this can be rewritten as

$$i' = T_o kA \left(\frac{1}{T}\frac{dT}{dx}\right)^2$$

$$= T_o kA \left(\frac{d \ln T}{dx}\right)^2$$

3-13 Using the results of Problem 3-12

(a) for the case of no cooling on the outer surface ($\dot{q}' \equiv 0$), show that the temperature variation in the rod is

$$T = T_1 + \left(\frac{T_2 - T_1}{L}\right)x$$

where L is the length of the rod. Also, show that the overall irreversibility for the rod, obtained by integration of i' over the rod length L, is

$$\dot{I} = kAT_o \left(\frac{T_1 - T_2}{L}\right)\left(\frac{1}{T_2} - \frac{1}{T_1}\right) \qquad (a)$$

Compare with the result of Example 3-5.

(b) Show that the temperature variation in the rod which *minimizes* its irreversibility is[7]

---

[7] This is an elementary problem in the *calculus of variations*. See for example F. B. Hildebrand, *Advanced Calculus for Applications*, 2nd ed., Prentice-Hall, Englewood Cliffs, New Jersey, 1976, Sec. 7-8.

$$T = T_1 \exp\left[\frac{x}{L} \ln\left(\frac{T_2}{T_1}\right)\right]$$

and the corresponding value for the overall irreversibility is

$$\dot{I}_{min} = \frac{kAT_o}{L}\left[\ln\left(\frac{T_2}{T_1}\right)\right]^2 \tag{b}$$

(c) Show that the ratio $\dot{I}/\dot{I}_{min}$ obtained by dividing Eq. (a) by Eq. (b) is

$$\frac{\dot{I}}{\dot{I}_{min}} = \frac{1}{\tau}\left[\frac{\tau - 1}{\ln \tau}\right]^2 \tag{c}$$

where $\tau = T_2/T_1$. For a discussion of Eq. (c) and an extension of the methodology of this problem applied to thermal insulation design, see A. Bejan, "A General Variational Principle for Thermal Insulation System Design," *Int. J. Heat Mass Transfer*, **22**, 1979, 219–227.

3-14 Determine the irreversibility in Btu/lb for an adiabatic throttling process ($h_i \approx h_e$) from 100 lbf/in.² to 20 lbf/in.² for
(a) steam entering as a saturated vapor.
(b) air as an ideal gas entering at 328°F.

3-15 Referring to Example 3-12, locate the control surface to include a portion of the immediate surroundings so heat transfer occurs at the temperature T_o. For this control volume, calculate on a unit mass basis the rate availability exits by heat transfer and the irreversibility. Compare with the results of Example 3-12. Discuss.

3-16 A well-insulated steam turbine at steady-state develops shaft work equal to 298 Btu/lb. At the inlet, $p_1 = 500$ lbf/in.², $T_1 = 900°F$. At the exit, $p_2 = 14.7$ lbf/in.². Determine the irreversibility in Btu/lb. Neglect changes in kinetic and potential energy.

3-17 Compute the *effectiveness* ϵ and *isentropic turbine efficiency* η_t for the turbine of Problem 3-16. (See Example 3-11 for definitions of ϵ and η_t.)

3-18 A well-insulated steam turbine at steady-state has steam entering at 1000 lbf/in.², 1100°F at a mass flow rate of 50,000 lb/hr. 20% of the flow is extracted at 300 lbf/in.² and 800°F and is diverted to a feedwater heater. The remainder continues through the turbine exiting at 1 lbf/in.² and quality $x = 99\%$. Determine the total shaft work output and the irreversibility in Btu/hr. Neglect changes in kinetic and potential energy.

3-19 A well-insulated steam turbine at steady-state can be operated at part-load conditions by throttling the steam to a lower pressure before it enters the turbine. Before throttling the steam is at 200 lbf/in.², 600°F. After throttling the pressure is 150 lbf/in.². The turbine exhaust pressure is fixed at 1 lbf/in.². Neglect changes in kinetic and potential energy.
(a) For the throttling valve, determine the irreversibility in Btu/lb.
(b) If the turbine effectiveness (see Example 3-11) is $\epsilon = 75\%$, determine its irreversibility in Btu/lb.

3-20 Air as an ideal gas flows steadily with negligible changes in kinetic and potential energy through the heater shown with an increase in temperature of 80°F. Electricity enters at a rate 0.1 kW. There are no stray heat transfers with the surroundings. See Fig. P3-20.

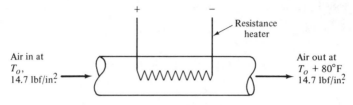

Figure P3-20

(a) Determine the rate availability enters with electricity in Btu/hr.
(b) Determine the increase in flow availability of the air in Btu/hr.
(c) Determine the irreversibility in Btu/hr.
(d) Dividing the result of part (b) by that of part (a), calculate a *second law efficiency* for the device. Discuss.

3-21 Saturated water vapor at 90 lbf/in.² enters a well-insulated mixing type of feedwater heater. Also entering is a stream of liquid water at 90 lbf/in.², 200°F. The combined stream exits as a saturated liquid at 90 lbf/in.². Determine the fraction of the exiting flow that enters as saturated water vapor. Also, determine the irreversibility in Btu/lb exiting the device. The heater operates at steady-state. There are no stray heat losses and kinetic and potential energies are negligible.

3-22 An ideal gas with constant specific heat c_p expands through a well-insulated turbine at steady-state. At the turbine inlet the state is defined by temperature T_i and pressure p_i. At the exit the pressure is p_e. Kinetic and potential energy changes are negligible.

(a) For an internally reversible expansion to p_e, show that the shaft work per unit mass flowing through is

$$\left(\frac{\dot{W}}{\dot{m}_f}\right)_{\text{int rev}} = \frac{kRT_i}{k-1}\left[1 - \left(\frac{p_e}{p_i}\right)^{(k-1)/k}\right] \quad \text{(a)}$$

where k is the specific heat ratio.

(b) For an actual expansion to p_e, show that the irreversibility per unit mass flowing through can be written in terms of T_o, gas constant R, specific heat ratio k, pressure ratio p_e/p_i, and the isentropic turbine efficiency η_t (η_t is the ratio of the actual work developed to the work given by Eq. (a)).

3-23 A counterflow heat exchanger operates at steady-state. On one side, saturated water vapor at 80°F enters and exits as a saturated liquid at 80°F. On the other side, liquid water enters at 60°F and exits at T_e. There are no stray heat transfers with the surroundings, pressure drops are negligible as are changes in kinetic and potential energy. The liquid side mass flow rate is 60 times the steam side mass flow rate. The liquid may be regarded as incompressible with specific heat, $c = 1$ Btu/lb°R.

(a) Find T_e.
(b) Determine the irreversibility of the overall heat exchanger in Btu/lb of steam.
(c) Show that the irreversibility of the condensing steam is zero.
(d) Show that the irreversibility of the heated liquid is zero.

Problems

(e) In view of the results of parts (c) and (d), explain why the answer to part (b) is nonzero.

3-24 On each side of the counterflow heat exchanger shown in Fig. P3-24 flows an ideal gas with constant specific heat c_p. The heat exchanger is at steady-state, there are no stray heat transfers with the surroundings, and kinetic and potential energy differences are negligible.

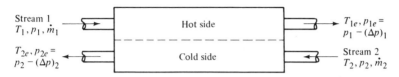

Figure P3-24

(a) Show that the irreversibility can be written as

$$\dot{I} = T_o \left[C_1 \ln\left(\frac{T_{1e}}{T_1}\right) + C_2 \ln\left(\frac{T_{2e}}{T_2}\right) \right. \\ \left. - C_1 \left(\frac{R}{c_p}\right)_1 \ln\left(1 - \frac{(\Delta p)_1}{p_1}\right) - C_2 \left(\frac{R}{c_p}\right)_2 \ln\left(1 - \frac{(\Delta p)_2}{p_2}\right) \right] \quad \text{(a)}$$

where $C_1 = \dot{m}_1 c_{p1}$, $C_2 = \dot{m}_2 c_{p2}$. It is assumed $C_1 < C_2$.

(b) Show that the exit temperatures T_{1e} and T_{2e} can be eliminated from Eq. (a) by combining the energy equation

$$C_1(T_1 - T_{1e}) + C_2(T_2 - T_{2e}) = 0$$

with the *heat exchanger effectiveness* ξ

$$\xi = \frac{T_{1e} - T_1}{T_2 - T_1}$$

(see for example, Frank Krieth, *Principles of Heat Transfer*, 3rd ed., Intext Educational Publishers, New York, 1973, Sec. 11-4) to give,

$$\dot{I} = T_o \left\{ C_1 \ln\left[1 + \xi\left(\frac{T_2}{T_1} - 1\right)\right] + C_2 \ln\left[1 - \frac{C_1}{C_2}\xi\left(1 - \frac{T_1}{T_2}\right)\right] \right. \\ \left. - C_1 \left(\frac{R}{c_p}\right)_1 \ln\left(1 - \frac{(\Delta p)_1}{p_1}\right) - C_2 \left(\frac{R}{c_p}\right)_2 \ln\left(1 - \frac{(\Delta p)_2}{p_2}\right) \right\} \quad \text{(b)}$$

Equation (b) shows that the irreversibility can be separated into a contribution associated with the stream-to-stream temperature difference ΔT (the first two terms) and a contribution associated with the frictional pressure drops (the last two terms).

In effect, the usual heat exchanger design procedure specifies the allowable ΔT and Δp values, thus dictating the irreversibility. In certain cases, the design process can be based on specification of the acceptable irreversibility level. The heat exchanger geometry is then determined to achieve this. Equation (b) plays an important role in this approach. For further detail, see A. Bejan, "The Concept of Irreversibility in Heat Exchanger Design: Counterflow Heat Exchangers for Gas-to-Gas Applications," *ASME J. Heat Trans.*, 99, August 1977, 374–380.

3-25 Figure P3-25 shows a system operating at steady-state for producing liquid Refrigerant 12. Vapor enters at flow port 1 and is cooled by saturated vapor being drawn from the insulated chamber. The fluid exiting the heat exchanger at location 2 is throttled to the pressure in the chamber. The chamber contains a two-phase liquid-vapor mixture at 0°F. Neglect all pressure drops except across the valve, neglect kinetic and potential energy changes, and assume there are no stray heat transfer effects with the surroundings.

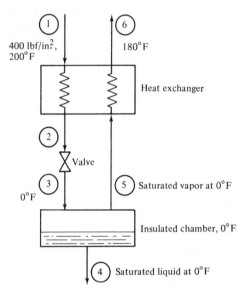

Figure P3-25

(a) Determine the amount of liquid drawn off at flow port 4 per lb of vapor entering at flow port 1.
(b) Calculate the irreversibility of the heat exchanger in Btu per lb of vapor entering at flow port 1.
(c) Calculate the irreversibility of the throttling valve in Btu per lb of vapor entering at flow port 1.

Further discussion related to the availability analyses of gas liquefaction systems is given in, C. R. Baker and R. L. Shaner, "A Study of the Efficiency of Hydrogen Liquefaction," *Int. J. Hydrogen Energy*, 3, 1978, 321–334.

3-26 A residential heat-pump system at steady-state is shown schematically in Fig. P3-26. Property data for the refrigerant are given in the accompanying table. Neglect flow kinetic and potential energy changes. Let $T_o = 32°F$, $p_o = 1$ atm.
(a) Locating the control surface to take in a portion of the immediate surroundings so that unavoidable heat transfer takes place at temperature T_o, determine the irreversibility in Btu/hr for the suction line 1–2 and for the compressor.
(b) Determine the change in flow availability experienced by the refrigerant through the condenser in Btu/hr. (A portion of this decrease is due to availability destruction within the condenser, the remainder is availability carried out by heat transfer.)

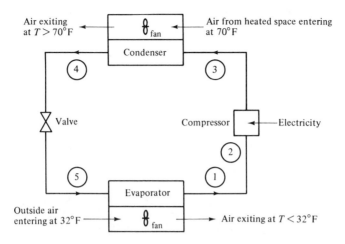

Figure P3-26

State	T (°F)	p (lbf/in.²)	h (Btu/lb)	s (Btu/lb°R)
1	24	41	96	0.2024
2	30	38	97.3	0.2064
3	200	219	123.1	0.2204
4	124	214.9	44.4	0.0875
5	21	42.8	44.4	0.0943
Dead state	32	14.7	99.4	0.2288

Refrigerant flow rate: 606 lb/hr
Electricity to compressor: 16,100 Btu/hr (4.72 kW)

(c) Determine the irreversibility of the throttling valve in Btu/hr.
(d) Determine the heat transfer of energy to the refrigerant in the evaporator in Btu/hr.
(e) Determine the change in flow availability experienced by the refrigerant through the evaporator in Btu/hr. Explain why the flow availability decreases even though there is a heat transfer to the refrigerant.
(f) Verify that the results of parts (a), (b), (c), and (e) account for all of the availability entering the cycle at the compressor with electricity.

For a more detailed discussion of this system see G. M. Reistad, "Availability Analysis of the Heating Process and A Heat-Pump System," *ASHRAE Symposium Paper*, LO-73-4, 1973, 29–35.

3-27 Show that the maximum work that can be extracted from a combined system consisting of an evacuated space with volume V and an environment at p_o is given by the product $p_o V$. Discuss this from the perspective of availability analysis.

3-28 Using an availability equation, determine the irreversibility in Btu for the filling process of Problem 2-21. Check your answer by means of $I = T_o \sigma$. *Hint:* When applying the availability equation, use the result of Problem 3-27.

3-29 Consider the steady *laminar* flow of an incompressible fluid through a horizontal tube with radius r_o. Let there be a uniform heat flux q_s around its circumference. The axial velocity profile for the fluid is

$$\mathcal{V}_x = \mathcal{V}_{\max}\left[1 - \left(\frac{r}{r_o}\right)^2\right] \qquad (a)$$

where

$$\mathcal{V}_{\max} = \frac{r_o^2}{4\mu}\left[-\frac{dp}{dx}\right]$$

and μ denotes the fluid viscosity. The temperature distribution for large (x/x_o) is

$$T - T' = \frac{q_s r_o}{k}\left[-4\left(\frac{x}{x_o}\right) - \left(\frac{r}{r_o}\right)^2 + \frac{1}{4}\left(\frac{r}{r_o}\right)^4 + \frac{7}{24}\right] \qquad (b)$$

where $x_o = \rho c_p \mathcal{V}_{\max} r_o^2 / k$, k is the fluid thermal conductivity and T' is the uniform fluid temperature at $x = 0$. (See, for example, R. B. Bird, W. E. Stewart, and E. N. Lightfoot, *Transport Phenomena* John Wiley & Sons, New York, 1960, Sec. 9.8.)

Equation (g) of Problem 2-22 expressed in cylindrical coordinates gives the entropy production *per unit of volume*

$$\sigma = \frac{k}{T^2}\left[\left(\frac{\partial T}{\partial x}\right)^2 + \left(\frac{\partial T}{\partial r}\right)^2\right] + \frac{\mu}{T}\left(\frac{\partial \mathcal{V}_x}{\partial r}\right)^2 \qquad (c)$$

(a) Neglecting axial conduction, combine Eqs. (a) to (c) to obtain,

$$\sigma = \frac{1}{k}\left\{\frac{q_s}{T}\left[\left(\frac{r}{r_o}\right)^3 - 2\left(\frac{r}{r_o}\right)\right]\right\}^2 + \frac{4\mu \mathcal{V}_{\max}^2 r^2}{T r_o^4} \qquad (d)$$

(b) The rate of entropy production *per unit of duct length*, σ', can be found by integration

$$\sigma' = 2\pi r_o^2 \int_0^1 \sigma \cdot \frac{r}{r_o} \cdot d\left(\frac{r}{r_o}\right)$$

Introducing Eq. (d), show that if the temperature variation over the tube cross-section is negligible compared with the absolute temperature

$$\sigma' = \frac{11(q')^2}{48\pi k T^2} + \frac{8\mu(\dot{m}_f)^2}{\pi T \rho^2 r_o^4} \qquad (e)$$

where $\dot{m}_f$ is the mass flow rate and q' is the heat transfer rate per unit of length: $q' = 2\pi r_o q_s$.

Since $i' = T_o \sigma'$, where i' is the irreversibility per unit of length, Eq. (e) shows once again that the irreversibility is determined by two effects: one due to heat transfer and the other to fluid friction. For a further discussion of Eq. (e) and additional applications of this methodology to convection heat transfer see A. Bejan, "A Study of Entropy Generation in Fundamental Convective Heat Transfer," *ASME J. Heat Trans.*, **101**, November 1979, 718–725.

Chapter 4

Second Law Efficiencies

The primary objective of this chapter is to introduce the concept of second law efficiency and to show its use in assessing the efficiency of energy utilization. An important secondary objective is to illustrate further the application of the availability principles developed in Chap. 3.

4-1 Introduction

Engineers make frequent use of efficiencies to gauge the performance of devices and processes. Many of these expressions are based on energy—they are first law efficiencies. Also useful are measures of performance that take into account limitations imposed by the second law. Efficiencies of this type are called here *second law efficiencies*.[1]

In this chapter, the energy and availability equations are used systematically to develop second law efficiency expressions for a number of cases. Additional cases are considered later in the book. These presentations are not meant to be exhaustive, nor are the second law efficiencies introduced with the idea that they ought to displace more familiar energy-based expressions, but rather with the objective of indicating their usefulness in gauging how effectively availability is used.

Illustration. To illustrate the idea of a performance parameter having to do with the second law and to contrast it with an analogous energy-based effi-

---

[1] Among several other terms used to identify second law efficiencies are *effectiveness, exergetic efficiency, rational efficiency,* and *thermodynamic efficiency ratio.*

ciency, consider a control volume at steady-state for which the energy and availability equations can be written, respectively, as

(Energy in) = (Energy out in product) + (Energy loss)
(Availability in) = (Availability out in product) + (Availability loss)
+ (Availability destruction)

In these equations, the term *product* might refer to shaft work or electricity developed, a certain heat transfer, some desired combination of heat and work, or possibly a particular exit stream (or streams). *Losses* are understood to include such things as waste heat and stack gases vented to the surroundings without use. The destruction term in the availability equation refers to availability destruction due to internal irreversibilities.

From either viewpoint, energy or availability, a gauge of how effectively the input is converted to the product is the ratio (product/input), that is

$$\eta = \frac{\text{Energy out in product}}{\text{Energy in}}$$
$$= 1 - \left(\frac{\text{Loss}}{\text{Input}}\right) \quad \text{(a)}$$

$$\epsilon = \frac{\text{Availability out in product}}{\text{Availability in}}$$
$$= 1 - \left[\frac{\text{Loss} + \text{destruction}}{\text{Input}}\right] \quad \text{(b)}$$

The second law efficiency ϵ frequently gives a finer understanding of performance than η. In computing η, the same weight is assigned to energy whether it be shaft work or a stream of low temperature fluid. Also, it centers attention on reducing "losses" to improve efficiency. The parameter ϵ weights energy flows by accounting for each in terms of availability. It stresses that *both* losses and internal irreversibilities need to be dealt with to improve performance. In many cases it is the irreversibilities that are more significant and the more difficult to deal with.

Equations (a) and (b) each define a *class* of efficiencies because a judgment has to be made about what is the product, what is counted as a loss, and even what is regarded as the input. Different decisions about these lead to different efficiency expressions within the class.

Other Efficiency Expressions. Other kinds of second law efficiency expressions also appear in the literature. One of these, appropriate only for devices at steady-state, is calculated as the ratio of the sum of the availability exiting to the sum of the availability entering. This is illustrated in Sec. 4-4. Another class of second law efficiencies is composed of *task* efficiencies.

4-1 Introduction

A task efficiency is the ratio of the *theoretical minimum input* required by the first and second laws to accomplish some task to the *actual input* for a particular means. In another form, the ratio of the actual output to the maximum theoretical output is evaluated. Table 4-1 gives examples. Additional examples are given later in the book.

The expressions introduced thus far represent only a sampling of the second law efficiencies which have appeared in the literature. There are several other forms that could be mentioned. Additional forms are brought into the text later. In using any of these expressions, care must be taken to understand fully how the parameter is defined and to avoid making comparisons with values computed with different definitions.

Discussion. One important use of second law efficiencies is to assess the performance of a device, plant, or industry relative to the average present-day level for like devices, plants, or industries. Another is to evaluate accurately steps taken to improve energy utilization. Some additional ideas about second law efficiencies and their use follow.

First, as suggested by the values of Table 4-1, the level of fuel consumption with current technology may be several times the minimum theoretical requirement. The gap which exists between the level of fuel consumption with current practices and the minimum theoretical requirement is a measure of the *potential for improvement*. Steps taken to reduce fuel consumption, to increase the second law efficiency, can result in the gap being diminished, but never closed.

This is because a second law efficiency of 100% is an unachievable goal. The theoretical limit could be attained only by conducting each step in a process

Table 4-1

Second Law (task) Efficiencies for Selected U.S. Industries in 1968[a]

$$\epsilon = \frac{\left(\begin{array}{c}\text{Theoretical minimum specific fuel consumption}\\ \text{based on availability analysis (Btu/ton)}\end{array}\right)}{(\text{Actual specific fuel consumption (Btu/ton)})}$$

Industry	ϵ(%)
Iron and steel	23
Petroleum refining	9
Primary aluminum production	13
Cement	10

[a] Based on data presented by E. P. Gyftopolous, L. J. Lazaridis, T. F. Widmer, *Potential Fuel Effectiveness*, Ballinger, Cambridge, Mass., 1974. Also, see E. H. Hall et. al., "Evaluation of the Theoretical Potential for Energy Conservation in Seven Basic Industries," NTIS Report PB-244772, U.S. Department of Commerce, Washington, DC, 1975.

without availability destructions or losses. Further, to execute such idealized processes would require in the limit an infinite time. Since all real processes are irreversible and require finite time for completion, the limit of 100% efficiency should not be regarded as a practical objective.

As a final point, it is total costs and not fuel consumption, or efficiencies, that dominate in the decision making process. An increase in efficiency to reduce fuel consumption, or otherwise utilize energy better, normally requires additional expenditures for facilities and operations. The tradeoff between fuel savings and additional investment needs to be weighed carefully.

The remainder of this chapter is devoted to the development and interpretation of second law efficiency expressions for a number of cases of practical interest. In each instance the expression is derived by a systematic use of the energy and availability equations.

4-2 Control Mass Applications

The purpose of this section is to develop a further understanding of the second law efficiency concept by considering three simplified cases. The first emphasizes that as one passes from fuel to end use, each step of the way exacts its tariff.

Example 4-1. A storage battery is charged with a certain amount of electricity W_e increasing its availability by ΔA. During discharge, the availability decreases in amount $-(\Delta A)$. Devise a second law efficiency for both the charging and discharging processes, and for the overall process of charging and discharging.

Solution. During the charging process, energy is added in amount W_e. Energy is also carried out in amount $Q_l (\geq 0)$ due to unavoidable heat transfer, the subscript l signifying "loss". The energy equation is

$$\Delta U = W_e - Q_l$$

In principle, the electricity flow can be converted entirely to shaft work in an idealized motor. In terms of availability, the worth of the electricity flow is identical to W_e. An availability equation for the charging process is

$$\Delta A = W_e - \int \left(1 - \frac{T_o}{T_l}\right) \delta Q_l - I_c \tag{a}$$

where I_c accounts for the destruction of availability due to irreversibilities within the battery during charging.

Equation (a) shows that the increase in availability is necessarily less than W_e because of heat transfer and internal irreversibilities. A second law efficiency in the form (increase/input) is

4-2 Control Mass Applications

$$\epsilon_c = \frac{\Delta A}{W_e}$$

In writing this, the heat transfer is regarded as a loss.

For the discharge process the availability equation is

$$(-\Delta A) = -\tilde{W}_e - \int \left(1 - \frac{T_o}{T_I}\right)\delta Q_I - I_d$$

$\tilde{W}_e$ denotes the electricity drawn out and I_d accounts for internal destruction during discharge.

Due to heat transfer and internal irreversibilities the electricity drawn out is less than the magnitude of the availability decrease. A second law efficiency can be written as

$$\epsilon_d = \frac{\tilde{W}_e}{\Delta A}$$

For the overall process of charging followed by discharging, an overall efficiency is the ratio of electricity out to electricity in, or

$$\epsilon_o = \frac{\tilde{W}_e}{W_e}$$

This can be expressed as the product of the efficiencies ϵ_c and ϵ_d, as

$$\epsilon_o = \frac{\tilde{W}_e}{W_e} = \left(\frac{\tilde{W}_e}{\Delta A}\right)\left(\frac{\Delta A}{W_e}\right)$$
$$= \epsilon_d \epsilon_c$$

Thus each step exacts its penalty. This reasoning can be continued as follows.

If it is assumed that the second law efficiency for the production and distribution of the electricity W_e is

$$\epsilon_p = \frac{W_e}{A_f}$$

where A_f is the availability of the fuel required (Chap. 7), an overall efficiency for the sequence from fuel input through battery discharge may be written as

$$\epsilon_o = \frac{\tilde{W}_e}{A_f} = \epsilon_d \epsilon_c \epsilon_p$$

In addition, the process by which some device makes use of the electricity $\tilde{W}_e$ brings in a further efficiency.

The following example illustrates the development of a second law efficiency for water heating and also shows the extremely wasteful nature of this common process.

Example 4-2. Devise a second law efficiency for the heating of water in a closed tank from temperature T_i to temperature $T_i + \Delta T$ using an electrical resistance heater. Discuss. Assume the water is incompressible.

Solution. During the water heating process, energy is added in amount W_e. Energy may also be carried out in amount $Q_l(\geq 0)$ due to unavoidable heat transfer. The energy equation is

$$\Delta U = W_e - Q_l$$

An energy-based efficiency in the form (increase/input) is $\eta = \Delta U/W_e$, which for an incompressible fluid is

$$\eta = \frac{mc\Delta T}{W_e} \tag{a}$$

where m is the mass of the water and c the specific heat.

The availability equation for the water heating process is

$$\Delta A = W_e - \int \left(1 - \frac{T_o}{T_l}\right)\delta Q_l - I$$

where I accounts for the destruction of availability due to internal irreversibilities and use is made of the fact that the availability input with the electricity is identical to W_e.

The increase in availability of the water is less than W_e due to heat transfer and internal irreversibilities. A second law efficiency in the form (increase/input) is

$$\epsilon = \frac{\Delta A}{W_e} \tag{b}$$

This is a *task* efficiency. Here the task is to heat water. The numerator is the minimum availability required to accomplish the task by *any* means whatever. The denominator is the availability input with the *particular* means under consideration. The key to achieving a high value for the efficiency is to match carefully the availability input with the intended use. This ideal is seldom realized in the ordinary practices of heating water.

Evaluating ΔA via Eq. (3-4a) for an incompressible fluid, Eq. (b) becomes

$$\epsilon = \frac{mc\{\Delta T - T_o \ln[(T_i + \Delta T)/T_i]\}}{W_e} \tag{c}$$

4-2 Control Mass Applications

Combining Eqs. (a) and (c),

$$\epsilon = \eta\left[1 - \frac{T_o}{\Delta T}\ln\left(\frac{T_i + \Delta T}{T_i}\right)\right]$$

Taking as sample values, $T_o = T_i = 520°R$ and $\Delta T = 60°R$

$$\epsilon \approx \eta/20$$

So even if the water heater had no heat loss ($\eta = 100\%$), the second law efficiency would be no more than 5% a value which shows the ordinary practice of water heating is extremely wasteful. The source of availability destruction here is that high potential electricity is used in a resistance device to produce relatively low temperature (low potential) warm water.

If the second law efficiency for the production and distribution of the electricity W_e is

$$\epsilon_p = \frac{W_e}{A_f}$$

where A_f is the availability of the fuel required (Chap. 7), an overall efficiency for the chain from power plant fuel to hot water is

$$\epsilon_o = \frac{\Delta A}{A_f} = \epsilon\epsilon_p$$

These results suggest that in view of the vast quantities of hot water generated daily, significant fuel savings might be realized by seeking a better match between input and use through waste heat recovery devices, solar heating systems, etc. This may well require more costly systems, however, and so the tradeoff between fuel savings and expenditures to obtain and maintain the needed equipment must be carefully evaluated.

The next example also stresses the importance of matching input to use by showing that for heat interactions the inefficiency of availability utilization is more pronounced the higher the temperature of the source is in relation to the temperature at which energy is delivered.

Example 4-3. A common process is the delivery of energy at a use temperature T_u from a source at temperature T_s. (a) Devise a second law efficiency for the case shown in Fig. E4-3. Assume a closed system at steady-state. (b) Evaluate the efficiency for a source temperature $T_s = 3500°F$, and use temperatures $T_u = 100°F$, $400°F$, and $800°F$ which are representative, respectively, of home heating, process steam generation, and industrial furnaces. Let $T_o = 60°F$.

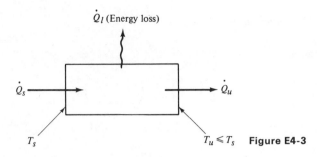

Figure E4-3

Solution. (a) Energy and availability rate equations for the steady closed system appear, respectively, as

$$0 = \dot{Q}_s - \dot{Q}_u - \dot{Q}_l$$

$$0 = \left(1 - \frac{T_o}{T_s}\right)\dot{Q}_s - \left(1 - \frac{T_o}{T_u}\right)\dot{Q}_u - \int_a \left(1 - \frac{T_o}{T_l}\right) q_l d\mathcal{Q} - \dot{I}$$

where $\dot{Q}_s$ and $\dot{Q}_u$ are positive in value. $\dot{Q}_l$ is a nonnegative number accounting for losses.

A second law efficiency in the form (product/input) is

$$\epsilon = \frac{\left(1 - \frac{T_o}{T_u}\right)\dot{Q}_u}{\left(1 - \frac{T_o}{T_s}\right)\dot{Q}_s}$$

This may be interpreted as the ratio of the work obtainable by passing $\dot{Q}_u$ to a reversible heat engine operating between T_u and T_o to the work obtainable by passing $\dot{Q}_s$ to a heat engine operating between T_s and T_o.

Introducing an energy-based efficiency in the form (product/input), $\eta = \dot{Q}_u/\dot{Q}_s$

$$\epsilon = \eta \left(\frac{1 - \frac{T_o}{T_u}}{1 - \frac{T_o}{T_s}}\right) \quad \text{(a)}$$

Equation (a) shows that a high efficiency η does not ensure that availability is being properly utilized. The temperatures T_u and T_s are also important, with availability utilization improving as the source temperature T_s approaches the utilization temperature T_u. This idea is emphasized by the sample calculations that follow.

(b) Letting $T_s = 3960°R$, $T_o = 520°R$, substitution into Eq. (a) results in

Application	$T_u(°R)$	ϵ/η
Home heating	560	0.082
Process steam generation	860	0.455
Industrial furnaces	1260	0.675

These values show that the closer the source temperature and use temperature, the better the match between availability input and intended use.

As a final point, it should be noted that when the temperature T_s is achieved by burning fuel there is a significant destruction of availability associated with the combustion process (Secs. 7-3 and 7-4). Accordingly, the overall efficiency for the sequence from fuel input to energy delivery at T_u would be less than that given by Eq. (a).

4-3 Cycle Applications

This section presents a sampling of second law efficiency expressions for power cycles and reversed cycles (heat pump and refrigeration cycles) that have appeared in the literature.

Power Cycles. Figure 4-1a shows the schematic of a power cycle. For illustrative purposes let energy be added to the cycle in heat interactions at temperatures T_A greater than T_R and rejected at temperatures T_R no lower than T_o.

Energy and availability equations for the cycle appear, respectively, as

$$W_{\text{cycle}} = Q_A - Q_R$$

$$0 = \int \left(1 - \frac{T_o}{T_A}\right) \delta Q_A - \int \left(1 - \frac{T_o}{T_R}\right) \delta Q_R - W_{\text{cycle}} - I$$

where $Q_A > 0$, $Q_R > 0$ and W_{cycle} is a positive number representing the *net* work developed by the cycle. I is the irreversibility. The first underlined term accounts for the availability added in a heat interaction at T_A. It can be interpreted as the work that could be developed by a reversible heat engine receiving energy at T_A and rejecting it at T_o. Similarly, the second underlined term can be viewed as the work that could be developed by a reversible heat engine receiving energy at T_R and rejecting it at T_o.

Regarding W_{cycle} as the product, a second law efficiency in the form (product/*net* supply) is

$$\epsilon = \frac{W_{\text{cycle}}}{\left[\int \left(1 - \frac{T_o}{T_A}\right) \delta Q_A - \int \left(1 - \frac{T_o}{T_R}\right) \delta Q_R\right]}$$

This expression does not charge as a loss the availability exiting in the heat interaction at T_R.

If the product of the cycle is identified as the net work developed and the availability exiting in the heat interaction at T_R is charged as a loss, a second law efficiency whose similarity to the *thermal efficiency* is apparent can be written as

$$\epsilon = \frac{W_{\text{cycle}}}{\int \left(1 - \frac{T_o}{T_A}\right) \delta Q_A}$$

This has the form (product/input) and compares the actual work to the work that could be developed by a reversible heat engine receiving energy at temperatures T_A and rejecting it at T_o. In terms of the thermal efficiency, $\eta = W_{\text{cycle}}/Q_A$

$$\epsilon = \eta \left[\frac{Q_A}{\int \left(1 - \frac{T_o}{T_A}\right) \delta Q_A} \right] \quad \text{(a)}$$

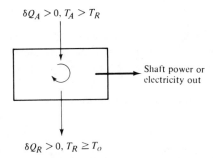

(a) Power cycle schematic. Energy added in a heat interaction at $T_A > T_R$ and rejected at $T_R \geq T_o$

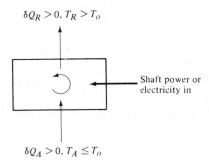

(b) Reversed cycle schematic. Energy added in a heat interaction at $T_A \leq T_o$ and rejected at $T_R \geq T_o$

Figure 4-1 Power and reversed cycle schematics.

Example 4-4. As shown in Fig. E4-4, a power cycle with a thermal efficiency $\eta = 30\%$ receives energy at $T_A = 1000°R$ from a TER at $T_s = 1200°R$ and rejects it at $T_R = 600°R$. For $T_o = 500°R$, apply Eq. (a) to each system shown.

4-3 Cycle Applications

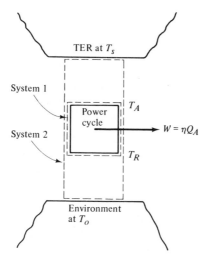

Figure E4-4

Solution. For System 1, Eq. (a) gives

$$\epsilon = \frac{\eta}{\left(1 - \dfrac{T_o}{T_A}\right)}$$

which is recognized as the ratio of the thermal efficiency of the given cycle to the thermal efficiency of a reversible heat engine operating between T_A and T_o. Inserting values, $\epsilon = 0.6$. This value accounts for the irreversibilities within the cycle and also charges it for the availability loss in the heat interaction at T_R.

For System 2, Eq. (a) gives

$$\epsilon = \frac{\eta}{\left(1 - \dfrac{T_o}{T_s}\right)}$$

This compares the actual thermal efficiency to that for a reversible heat engine operating between T_s and T_o. The second law efficiency given by this expression includes the effect of irreversibilities within the cycle as well as the effect of the irreversibilities associated with the heat transfer between TER and cycle, and between cycle and environment. Inserting values, $\epsilon = 0.51$.

Reversed Cycles. Figure 4-1*b* shows the schematic of a reversed cycle. For illustrative purposes let energy be added to the cycle in a heat interaction at temperatures that do not exceed T_o and rejected at temperatures no lower than T_o.

Energy and availability equations for the cycle appear, respectively, as

$$W_{\text{cycle}} = Q_R - Q_A$$

$$0 = \int \left(1 - \frac{T_o}{T_A}\right)\delta Q_A - \int \left(1 - \frac{T_o}{T_R}\right)\delta Q_R + W_{\text{cycle}} - I$$
$\underline{\phantom{\int \left(1 - \frac{T_o}{T_A}\right)\delta Q_A}}$

where $Q_R > 0$, $Q_A > 0$ and W_{cycle} is a *positive* number representing the net work done *on* the cycle.

If energy is added at $T_A < T_o$, the underlined term in the availability equation is negative. As noted in Sec. 3-5, when a heat interaction takes place at temperatures less than T_o, it is the source that gains availability and the system receiving energy that loses it. This point is emphasized by rearranging the availability equation into the form

$$\text{Input} = \text{Output} + \text{destruction}$$

or

$$W_{\text{cycle}} = \left[\int \left(1 - \frac{T_o}{T_R}\right)\delta Q_R\right] + \left[-\int \left(1 - \frac{T_o}{T_A}\right)\delta Q_A\right] + I$$

A second law efficiency in the form (output/input) is

$$\epsilon' = \frac{\left[\int \left(1 - \frac{T_o}{T_R}\right)\delta Q_R\right] + \left[-\int \left(1 - \frac{T_o}{T_A}\right)\delta Q_A\right]}{W_{\text{cycle}}}$$

By distinguishing between heat pump and refrigeration cycles, other expressions can be devised.

The objective of a *heat pump* is to deliver energy at an elevated temperature. If this is identified as the product, a second law efficiency in the form (product/input) is

$$\epsilon = \frac{\int \left(1 - \frac{T_o}{T_R}\right)\delta Q_R}{W_{\text{cycle}}}$$

Expressed in terms of the *coefficient of performance*, $\beta = Q_R/W_{\text{cycle}}$

$$\epsilon = \beta \left[\frac{\int \left(1 - \frac{T_o}{T_R}\right)\delta Q_R}{Q_R}\right] \tag{b}$$

Example 4-5. Consider a home electric heat pump with a coefficient of performance of 3.6. If the device delivers energy at 120°F, determine its second law efficiency as given by Eq. (b). Let $T_o = 490°R$.

Solution. Since energy is delivered at a constant temperature, Eq. (b) reduces to

$$\epsilon = \beta\left(1 - \frac{T_o}{T_R}\right)$$

Inserting numerical values: $\beta = 3.6$, $T_o = 490°R$, $T_R = 580°R$, $\epsilon = 56\%$.

If it is assumed that the second law efficiency for the production and distribution of the electricity required to drive the device is

$$\epsilon_p = \frac{W_{cycle}}{A_f} = 37\%$$

where A_f is the availability of the fuel required (Chap. 7), an overall efficiency is

$$\epsilon_o = \frac{Q_R\left(1 - \frac{T_o}{T_R}\right)}{A_f} = \epsilon_p \epsilon$$

$$= 21\%$$

By way of comparison, if the home is heated by a furnace burning a fossil fuel, the second law efficiency is on the order of 12% (Example 7-3).

Identifying the objective of the refrigeration cycle to be the maintenance of a cold region, a second law efficiency in the form (product/input) is

$$\epsilon = \frac{\left[-\int \left(1 - \frac{T_o}{T_A}\right) \delta Q_A\right]}{W_{cycle}}$$

Expressed in terms of the *coefficient of performance*, $\beta_R = Q_A/W_{cycle}$

$$\epsilon = \beta_R \left\{\frac{\left[-\int \left(1 - \frac{T_o}{T_A}\right) \delta Q_A\right]}{Q_A}\right\}$$

4-4 Control Volume Applications

In this section second law efficiency expressions are presented for a number of important cases involving control volumes at steady-state.

Turbines, Pumps, and Compressors. Assuming steady-state and negligible kinetic and potential energy changes, the energy and availability rate equations on a unit mass basis are, respectively,

$$0 = \frac{\dot{Q}}{\dot{m}_f} - \frac{\dot{W}}{\dot{m}_f} + (h_i - h_e)$$

$$0 = \frac{1}{\dot{m}_f} \int_a \left(1 - \frac{T_o}{T_s}\right) q_s d\mathcal{Q} - \frac{\dot{W}}{\dot{m}_f} + a_{fi} - a_{fe} - \frac{\dot{I}}{\dot{m}_f}$$

The turbine *effectiveness* evaluates how effectively the turbine converts the flow availability decrease to the product: work. That is

$$\epsilon = \frac{\dot{W}/\dot{m}_f}{(a_{fi} - a_{fe})}$$

This assumes the heat transfer of energy is lost to the surroundings. Applications of the turbine effectiveness are found in Examples 3-11 and 3-12.

For a pump or compressor, the corresponding efficiency appears as

$$\epsilon = \frac{a_{fe} - a_{fi}}{(-\dot{W}/\dot{m}_f)}$$

Throttling Process. Assuming steady, adiabatic flow and negligible kinetic and potential energy changes, the energy and availability rate equations on a unit mass basis are, respectively,

$$0 = h_i - h_e$$
$$0 = a_{fi} - a_{fe} - \dot{I}/\dot{m}_f$$

There being no "product," a second law efficiency can be devised in the form (output/input) as

$$\epsilon' = \frac{a_{fe}}{a_{fi}} = 1 - \frac{\dot{I}/\dot{m}_f}{a_{fi}}$$

This is a parameter that measures the irreversibility. Applying it to the case of Example 3-8 gives $\epsilon' = 77\%$.[2]

Heat Exchange with Mixing. An open feedwater heater is shown in Fig. 4-2. For simplicity, it is assumed the device is adiabatic and changes in kinetic and potential energies are negligible.

The energy and availability rate equations written per unit mass of mixture are, respectively

$$0 = yh_1 + (1-y)h_2 - h_3$$
$$0 = ya_{f1} + (1-y)a_{f2} - a_{f3} - \dot{I}/\dot{m}_3$$

where $y = \dot{m}_1/\dot{m}_3$.

[2] For additional discussion see Example 6-1.

4-4 Control Volume Applications

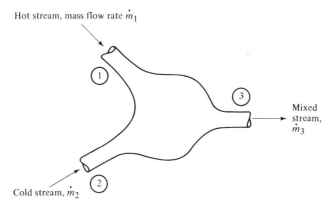

Figure 4-2 Open heater.

The availability rate equation may be rearranged as

$$0 = y(a_{f1} - a_{f3}) - (1 - y)(a_{f3} - a_{f2}) - \dot{I}/\dot{m}_3$$

Regarding the cold stream to be heated at the expense of the hot stream, the product is identified as the increase of availability of the cold stream: $(1 - y)(a_{f3} - a_{f2})$, and $y(a_{f1} - a_{f3})$ may be taken as the availability supplied. Then

$$\epsilon = \frac{(1 - y)(a_{f3} - a_{f2})}{y(a_{f1} - a_{f3})}$$

Applying this to the case of Example 3-9, $\epsilon = 44\%$.

If the exchange of energy occurs below the temperature of the environment, T_o, it is the hot stream that gains availability and the cold stream that loses it (Sec. 3-5). Accordingly, the efficiency ϵ must be written to reflect this.

Another efficiency parameter, in the form (output/input), is

$$\epsilon' = \frac{a_{f3}}{y a_{f1} + (1 - y) a_{f2}}$$

Heat Exchange without Mixing. The heat exchanger shown in Fig. 4-3 is assumed to operate adiabatically and at steady-state, changes in kinetic and potential energy are neglected. The energy and availability rate equations take the form, respectively,

$$0 = \dot{m}_h[h_1 - h_2] - \dot{m}_c[h_4 - h_3]$$
$$0 = \dot{m}_h[a_{f1} - a_{f2}] - \dot{m}_c[a_{f4} - a_{f3}] - \dot{I}$$

where $\dot{m}_h$ is the mass flow rate of the heating stream and $\dot{m}_c$ is that for the heated stream.

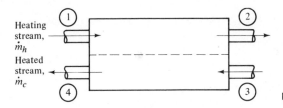

Figure 4-3 Heat exchanger.

If the product is taken as the increase in availability of the heated stream ($\dot{m}_c(a_{f4} - a_{f3})$) and the availability supplied as the change in availability of the heating stream ($\dot{m}_h(a_{f1} - a_{f2})$)

$$\epsilon = \frac{\dot{m}_c(a_{f4} - a_{f3})}{\dot{m}_h(a_{f1} - a_{f2})}$$

For heat exchange below the environmental temperature, T_o, it is the heating stream that gains the availability and the heated stream that loses it (Sec. 3-5); in such cases, the efficiency must be rewritten to reflect this.

An efficiency in the form (output/input) is

$$\epsilon' = \frac{\dot{m}_h a_{f2} + \dot{m}_c a_{f4}}{\dot{m}_h a_{f1} + \dot{m}_c a_{f3}}$$

Discussion. Though both efficiency expressions, ϵ and ϵ', appear in the literature, the efficiency ϵ is often preferred since it evaluates how effectively the input, or supply, is converted to the product. The parameter ϵ' can be used to make an evaluation in cases where a product is difficult, or impossible, to identify (throttling process).

4-5 Closure

The second law efficiency expressions presented in this chapter are representative of those that have appeared in the literature. No attempt has been made to be exhaustive. However, further elaboration on the second law efficiency concept is provided in the problems included with this chapter as well as in several of the end of chapter problems included with Chaps. 7, 8, and 9. In addition, the concept is discussed and illustrated in a number of places in the remainder of the text.

Selected References

1. BOSNJAKOVIC, F., *Technical Thermodynamics*, (P.L. Blackshear, Jr., translator), Holt-Rinehart-Winston, New York, 1965.
2. KOTAS, T. J., "Exergy Criteria of Performance for Thermal Plant" *Int. J. Heat & Fluid Flow*, **2**, 4, 1980, 147–163.

3. REISTAD, G. M., "Available Energy Conversion and Utilization in the United States," *Trans. ASME, J. Engr. Power*, **97**, July 1975, 429–434.
4. WOLFE, H. C. (Ed.), *Efficient Use of Energy*, American Institute of Physics Conference Proceedings No. 25, American Institute of Physics, New York, 1975.

Problems

4-1 Devise and evaluate a second law efficiency for the gear box of Problem 2-14. Let $T_o = 520°R$.

4-2 Devise and evaluate a second law efficiency for the gas turbine power plant of Example 3-7. Explain how the efficiency value would change if based on the fuel input to the plant.

4-3 Calculate the effectiveness for each of the two turbine stages of Problem 3-18.

4-4 Calculate the efficiency ϵ for the feedwater heater of Problem 3-21.

4-5 Calculate the efficiency ϵ for the heat exchanger of Problem 3-23.

4-6 Calculate the efficiency ϵ for a counterflow heat exchanger operating at steady-state as follows. On one side air enters at 850°R, 60 lbf/in.² and exits at 1000°R, 50 lbf/in.². On the other side, air enters at 1300°R, 16 lbf/in.² and exits at T_e, 14.7 lbf/in.². The mass flow rates are equal on both sides, there are no stray heat transfers, and kinetic and potential energy changes are negligible. Assume the ideal gas model with $c_p = 0.24$ Btu/lb°R. Let $T_o = 520°R$, $p_o = 14.7$ lbf/in.²

4-7 Steam enters an adiabatic nozzle at 60 lbf/in.², 500°F with negligible velocity. It expands steadily to 14.7 lbf/in.², 250°F with a negligible change in potential energy. Devise and evaluate a second law efficiency for the nozzle. Let $T_o = 520°R$, $p_o = 14.7$ lbf/in.²

4-8 Three devices are under consideration for delivering a heat transfer of energy at a use temperature $T_u = 140°F$. In each case, the energy *loss* is 10% of the supply. The devices, shown schematically in Fig. P4-8, are as follows:
(a) A solar collector system which supplies energy at $T_s = 240°F$.
(b) A more elaborate solar system which supplies energy at $T_s = 340°F$.
(c) An electrical resistance device which supplies energy at $T_s = 540°F$.
Using the approach of Example 4-3, devise and evaluate a second law efficiency

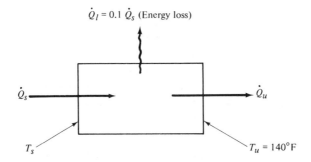

Figure P4-8

for each device. Discuss your results from the viewpoint of matching the input with the intended use. Let $T_o = 40°F$.

4-9 A steam ejector is shown in Fig. P4-9. Assuming the product is the increase in availability of the fluid entering at 1, devise a second law efficiency for the device which operates at steady-state with no stray heat transfers and negligible changes in kinetic and potential energies.

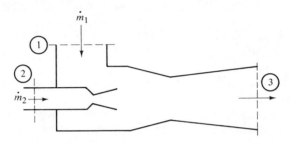

Figure P4-9

4-10 As shown in Fig. P4-10, an absorption type heat pump is *driven* by a high temperature heat source and a small input of electricity. The source delivers energy to the heat pump cycle at $T_s > T_o$. The cycle delivers energy to the heated space at a uniform temperature $T_H > T_o$. Energy is drawn into the cycle from the surroundings which are at T_o. Devise a second law efficiency for the device.

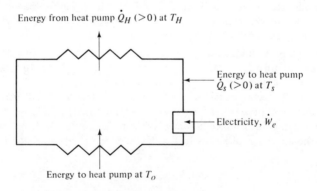

Figure P4-10

4-11 Figure P4-11 shows a control volume encompassing a turbine. Because of the way the control surface is selected there are $(n + 1)$ inlets and $(n + 1)$ exits.

Assuming no losses of availability due to heat transfer, the second law efficiency ϵ' in the form (output/input) is

$$\epsilon' = \frac{(\dot{W}_t/\dot{m}_f) + a_{fe} + \sum_{k=1}^{n} a_{fe,k}}{a_{fi} + \sum_{j=1}^{n} a_{fi,j}} \tag{a}$$

Problems

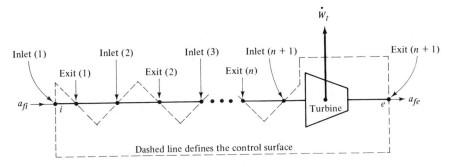

Figure P4-11

where the subscript k identifies all exits except the one downstream of the turbine which is identified as a_{fe}, and the subscript j identifies all inlets except the first which is identified as a_{fi}.

If there are no irreversibilities anywhere except across the turbine, show that Eq. (a) can be written as

$$\epsilon' = \frac{(\dot{W}_t/\dot{m}_f) + a_{fe} + na_{fi}}{(n+1)a_{fi}} \qquad \text{(b)}$$

$$= \frac{\left(1 - \dfrac{\dot{i}}{\dot{m}_f a_{fi}}\right)}{(n+1)} + \left(\frac{n}{n+1}\right)$$

where $\dot{i}$ accounts for the irreversibility across the turbine. Use Eq. (b) to show that in the limit as $n \longrightarrow \infty$, $\epsilon' \longrightarrow 1$. That is, an "efficiency" of 100% is achieved without taking steps to reduce the turbine irreversibility. Discuss.

4-12 On each side of the counterflow heat exchanger shown in Fig. P4-12 an ideal gas with constant c_p flows. The device is at steady-state, there are no stray heat transfers, the pressure drop on both sides is negligible, as are kinetic and potential energy changes.

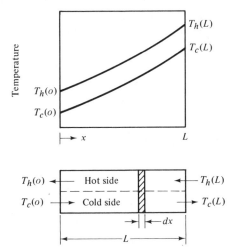

Figure P4-12

(a) Show that the efficiency ϵ can be written as

$$\epsilon = \frac{[T_c(L) - T_c(o)] - T_o \ln\left(\frac{T_c(L)}{T_c(o)}\right)}{(C_h/C_c)\left[(T_h(L) - T_h(o)) - T_o \ln\left(\frac{T_h(L)}{T_h(o)}\right)\right]}$$

where $C_h = \dot{m}_h c_{ph}$, $C_c = \dot{m}_c c_{pc}$.

(b) Using the same approach as in part (a) but applied to a differential length dx, show that a *local* efficiency is given by

$$\epsilon(x) = \frac{T_h(x)[T_c(x) - T_o]}{T_c(x)[T_h(x) - T_o]} \quad (a)$$

(c) Verify that at any location x

$$T_h(x) - T_c(x) = [T_h(o) - T_c(o)]Z \quad (b)$$

where

$$Z = \exp\left[Uw\left(\frac{1}{C_h} - \frac{1}{C_c}\right)x\right]$$

U is an *overall unit conductance* (assumed constant) and w is the width of the heat transfer area.

(d) Use Eq. (b) to obtain the temperature variations

$$T_h(x) = T_h(o) + \left[\frac{T_h(o) - T_c(o)}{1 - (C_h/C_c)}\right](Z - 1)$$

$$T_c(x) = T_c(o) + \left[\frac{T_h(o) - T_c(o)}{(C_c/C_h) - 1}\right](Z - 1) \quad (c)$$

(e) Finally, using Eqs. (c) show that Eq. (a) can be expressed in the form

$$\epsilon(x) = \frac{(Z + a)(Z + b)}{(Z + c)(Z + d)}$$

where a, b, c, and d are combinations of the parameters involved: $T_h(o)$, $T_c(o)$, C_h/C_c, and T_o.

For further discussion of the local second law efficiency $\epsilon(x)$ see P. J. Golem and T. A. Brzustowski, "Second-Law Analysis of Energy Processes, Part II: The Performance of Simple Heat Exchangers," *Trans. Can. Soc. Mech. Eng.* **4**, 4, 1976–1977, 219–226.

Chapter 5

Availability Property Relations

This chapter presents availability and flow availability relations for pure substances and gas mixtures. There is considerably more stress on the evaluation of flow availability than on availability due to its importance in the analysis of control volumes. Property relations for gases and vapors are emphasized over those for liquids.

5-1 Introduction

The specific availability a and the specific flow availability a_f are, respectively,

$$a = (u - u_o) + p_o(v - v_o) - T_o(s - s_o) \qquad \text{(5-1a)}$$
$$a_f = (h - h_o) - T_o(s - s_o) \qquad \text{(5-1b)}$$

In writing these expressions, kinetic and potential energies have been omitted. Moreover, it is understood that the dead state being used is the restricted dead state (Sec. 3-3).

The objective of this chapter is to discuss various means for evaluating the availability functions, including graphical methods. The presentation centers on the evaluation of availability and flow availability for pure substances and gas mixtures. Further aspects of property evaluation, including those associated with multicomponent systems are taken up in Chaps. 6 and 7.

In performing availability analyses it is necessary to evaluate the availability and/or flow availability functions at various states. Examples 5-1 and 5-2 review the use of property tables for this purpose.

Example 5-1. Determine the specific flow availability of nitrogen at 1000 lbf/in.^{2}, 300°R relative to the restricted dead state defined by $T_o = 520°R$, $p_o = 14.7$ lbf/in.^{2}. Neglect kinetic and potential energies.

Solution. Using Appendix Table B-4 for nitrogen, at 1000 lbf/in.^{2}, 300°R, $h = 115.22$ Btu/lb, $s = 0.5514$ Btu/lb°R; at 14.7 lbf/in.^{2}, 520°R, $h_o = 193.72$, $s_o = 1.0464$. Then

$$a_f = h - h_o - T_o(s - s_o)$$
$$= (115.22 - 193.72) - 520(0.5514 - 1.0464)$$
$$= -78.5 + 257.4$$
$$= 178.9 \text{ Btu/lb } (5012 \text{ Btu/lbmol})$$

The two availability functions are related as follows (Sec. 3-6)

$$a = a_f + v(p_o - p) \tag{5-1c}$$

Equation (5-1c) is valid for both gases and liquids. It can be recast in a form that is useful primarily for gases by introducing the compressibility factor, $Z = pv/RT$ (Sec. 2-4)

$$a = a_f + RTZ\left(\frac{p_o}{p} - 1\right) \tag{5-1d}$$

It will be recalled that Z can be evaluated from a compressibility chart for a wide range of gases and vapors.

Example 5-2. Determine the specific availability of nitrogen for the conditions of Example 5-1 using (a) property tables and Eq. (5-1c), and (b) the compressibility chart and Eq. (5-1d).

Solution. (a) Using Appendix Table B-4 for nitrogen at 1000 lbf/in.^{2}, 300°R, $v = 0.0828$ ft^3/lb. With Eq. (5-1c) and a_f from Example 5-1

$$a = 178.9 + (0.0828)(14.7 - 1000)\left(\frac{144}{778}\right)$$
$$= 163.8 \text{ Btu/lb}$$

(b) For nitrogen, $T_c = 227.2°R$, $p_c = 493$ lbf/in.^{2}. The reduced pressure p_r and reduced temperature T_r are

$$p_r = \frac{p}{p_c} = \frac{1000}{492} = 2.03$$

$$T_r = \frac{T}{T_c} = \frac{300}{227.1} = 1.32$$

5-2 Availability Property Diagrams

Using Appendix Fig. B-1, Z is about 0.73. Substituting into Eq. (5-1d),

$$a = 178.9 + \frac{(1.986)(300)(0.73)}{28.02}\left(\frac{14.7}{1000} - 1\right)$$
$$= 163.6 \text{ Btu/lb}$$

In the remainder of this chapter the evaluation of a_f is stressed because it is the important parameter for analysis of control volumes. Moreover, as shown by Eqs. (5-1c and d), the specific availability is readily determined from a_f. In Sec. 5-2, property diagrams for certain important substances are discussed. In Sec. 5-3, ideal gas relations are featured. Finally, the use of generalized property charts is taken up in Sec. 5-4.

5-2 Availability Property Diagrams

Graphical displays of property relations are often convenient for engineering analysis. Familiar examples are Mollier diagrams for steam turbine processes and psychrometric charts for heating, ventilating, and air conditioning (HVAC) applications. Diagrams giving flow availability or availability in terms of other key properties can also be useful.

Detailed availability property diagrams can aid analysis, but they have one noteworthy shortcoming. Since availability varies with the dead state, diagrams constructed for one dead state cannot in general be used with accuracy for any other dead state. A similar difficulty exists with the ordinary psychrometric chart, normally tied to a mixture pressure of one atmosphere. An availability diagram keyed to a particular dead state frequently can be useful in a qualitative sense even when another dead state is being used; but of course calculations should be based on tabulated data, other more exact graphical property data, or when applicable the methods presented in Secs. 5-3 and 5-4.

Figure 5-1 shows the schematic of a diagram giving the quantity $b = h - T_o s$ versus entropy for water.[1] A figure of this type is especially useful for the analysis of steam turbine processes. The work for adiabatic operation at steady-state with negligible changes in kinetic and potential energy, $\dot{W}/\dot{m}_f = h_i - h_e$, is readily evaluated from the figure. The irreversibility, $\dot{I}/\dot{m}_f = T_o(s_e - s_i)$, can be determined using the abscissa. And using the ordinate, the turbine *effectiveness* ϵ is easily obtained as

$$\epsilon = \frac{\dot{W}/\dot{m}_f}{b_i - b_e}$$

[1] For a detailed figure see Ref. [1]. On this chart $p_o = 1$ atm and T_o ranges in ten-degree intervals from 30°F to 70°F.

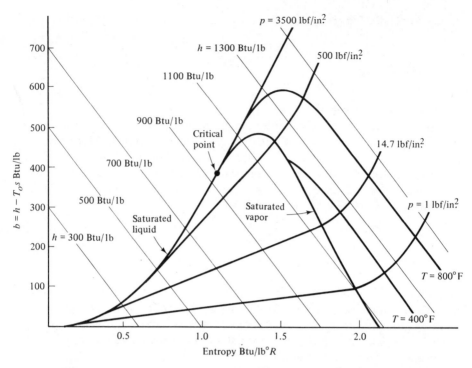

Figure 5-1 $b = (h - T_o s)$ versus entropy for water (p_o = 1 atm, T_o ranges in ten degree intervals from 30°F to 70°F).

The schematic of a flow availability versus enthalpy diagram for Refrigerant 12 is shown in Fig. 5-2.[2] This type of figure is convenient for analysis of the vapor-compression refrigeration cycle. For example, it enables the work of the adiabatic compressor to be determined without difficulty, $\dot{W}/\dot{m}_f = h_i - h_e$. Also, both the numerator and denominator of the second law efficiency ϵ for the compressor (Sec. 4-4) are read directly from the diagram

$$\epsilon = \frac{a_{fe} - a_{fi}}{h_e - h_i}$$

The second law efficiency of the overall cycle is readily determined by use of the figure. In addition, the throttling process is conveniently shown and its irreversibility also may be read directly from the diagram.

Availability property diagrams in terms of other coordinates have also appeared in the literature. As examples, Fig. 5-3a is the schematic of a diagram giving flow availability versus temperature for water and a plot for nitrogen in terms of the same coordinates is shown schematically in Fig. 5-3b.[3]

---

[2] For a detailed figure based on $p_o = 1$ atm, $T_o = 77°F$, see Ref. [3]. A chart for helium in terms of the same coordinates, based on $p_o = 1$ atm, $T_o = 300$ K, is given in Ref. [5].

[3] For detailed figures based on $p_o = 1$ atm, $T_o = 298$ K, see Ref. [2], pp. 228, 231.

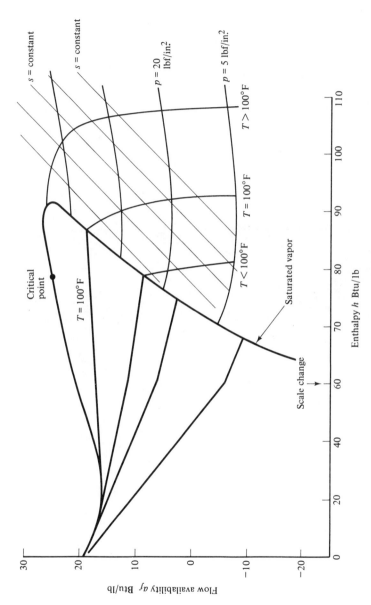

Figure 5-2 Flow availability $a_f = [h - h_o - T_o(s - s_o)]$ of Refrigerant 12 ($T_o = 77°F$, $p_o = 14.7$ lbf/in.²).

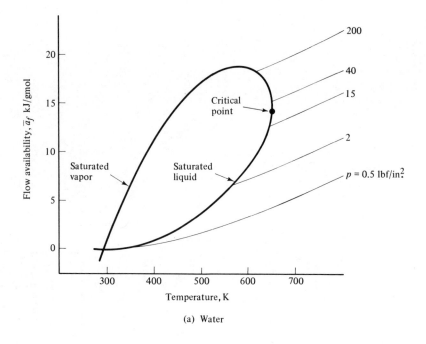

(a) Water

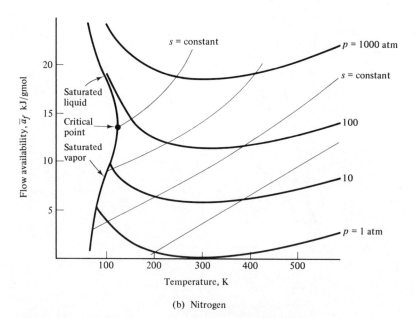

(b) Nitrogen

Figure 5-3 Flow availability $\bar{a}_f = [\bar{h} - \bar{h}_o - T_o(\bar{s} - \bar{s}_o)]$ versus temperature ($T_o = 298$ K, $p_o = 1$ atm).

5-3 Ideal Gas Relations

This section presents means for evaluating the availability functions a and a_f utilizing the ideal gas model. Tabular, graphical, and computational approaches are considered.

To evaluate the flow availability requires the determination of both enthalpy and entropy. For an ideal gas

$$\bar{h}^* - \bar{h}_o^* = \int_{T_o}^{T} \bar{c}_p(T)dT$$

$$\bar{s}^* - \bar{s}_o^* = \int_{T_o}^{T} \frac{\bar{c}_p(T)}{T} dT - \bar{R} \ln(p/p_o)$$

where the asterisk* denotes the ideal gas. Introducing these into Eq. (5-1b)

$$\bar{a}_f^*(T, p) = \int_{T_o}^{T} \left(1 - \frac{T_o}{T}\right) \bar{c}_p(T)dT + \bar{R}T_o \ln(p/p_o)$$

The last result is shown in a convenient form as

$$\bar{a}_f^*(T, p) = \bar{a}_f^*(T, p_o) + \bar{R}T_o \ln(p/p_o) \tag{5-3a}$$

where

$$\bar{a}_f^*(T, p_o) = \int_{T_o}^{T} \left(1 - \frac{T_o}{T}\right) \bar{c}_p(T)dT \tag{5-3b}$$

The remainder of this section concerns the evaluation of $\bar{a}_f^*(T, p_o)$ for pure ideal gases and ideal gas mixtures via ideal gas tables, direct computation, and when the specific heat ratio is constant, graphical means.

Evaluation of $\bar{a}_f^*(T, p_o)$ by Use of Ideal Gas Tables. Appendix Table B-6 presents ideal gas properties for a number of important gases. Evaluation of $\bar{a}_f^*(T, p_o)$ is easily accomplished by use of these tables by noting that Eq. (5-3b) may be expressed in terms of table variables $\bar{h}^*(T)$ and $\bar{s}'(T)$ as

$$\bar{a}_f^*(T, p_o) = [\bar{h}^*(T) - \bar{h}^*(T_o)] - T_o[\bar{s}'(T) - \bar{s}'(T_o)] \tag{5-3c}$$

Example 5-3. Repeat Example 5-1 but use the ideal gas table, Appendix Table B-6B. Compare the result with that of Example 5-1.

Solution. From Appendix Table B-6B, at 300°R, $\bar{h}^* = 2082$ Btu/lbmol, $\bar{s}' = 41.695$ Btu/lbmol°R; at 520°R, $\bar{h}^* = 3611.3$, and $\bar{s}' = 45.519$. With Eq. (5-3c),

$$\bar{a}_f^*(300°R, p_o) = (2082 - 3611.3) - 520(41.695 - 45.519)$$
$$= -1529.3 + 1988.48$$
$$= 459.18 \text{ Btu/lbmol}$$

Then, with Eq. (5-3a)

$$\bar{a}_f^*(300°R, 1000 \text{ lbf/in.}^2) = 459.18 + (1.986)(520)\ln(1000/14.7)$$
$$= 459.18 + 4357.98$$
$$= 4817.2 \text{ Btu/lbmol } (172 \text{ Btu/lb})$$

In this case, agreement with the value determined from the real gas property tables is obtained to within 5%. Further discussion of this point is provided in Example 5-5.

Evaluating $\bar{a}_f^*(T, p_o)$ by Direct Computation. By using a suitable analytical expression for the specific heat $\bar{c}_p$, the integral of Eq. (5-3b) can be performed. This approach is illustrated by use of the following convenient form

$$\bar{c}_p(T) = A + BT + CT^2 + DT^3 \tag{5-3d}$$

where the coefficients A, B, C, and D are given in Appendix Table B-5 for several important substances.

Introduction of Eq. (5-3d) into Eq. (5-3b) and performing the integral results in

$$\bar{a}_f^*(T, p_o) = (A - T_o B)[T - T_o] + \left(\frac{B - T_o C}{2}\right)[T^2 - T_o^2]$$
$$+ \left(\frac{C - T_o D}{3}\right)[T^3 - T_o^3] + \frac{D}{4}[T^4 - T_o^4] - T_o A \ln\left(\frac{T}{T_o}\right) \tag{5-3e}$$

Evaluation of $\bar{a}_f^*(T, p_o)$ for Constant Specific Heat Ratio. In some instances the specific heat is constant, or approximately so, over the temperature range under consideration. Then, Eq. (5-3b) gives

$$\bar{a}_f^*(T, p_o) = \bar{c}_p\left[(T - T_o) - T_o \ln\left(\frac{T}{T_o}\right)\right]$$

In dimensionless form this appears as

$$\frac{\bar{a}_f^*(T, p_o)}{\bar{c}_p T_o} = \frac{T}{T_o} - 1 - \ln\left(\frac{T}{T_o}\right)$$
$$= F\left(\frac{T}{T_o}\right)$$

where $F(x)$ is defined as $x - 1 - \ln x$.

5-3 Ideal Gas Relations

Combining this with Eq. (5-3a) and using the relation $\bar{c}_p = k\bar{R}/(k-1)$, gives the *reduced* flow availability

$$\frac{\bar{a}_f^*(T, p)}{\bar{c}_p T_o} = F\left(\frac{T}{T_o}\right) + \ln\left(\frac{p}{p_o}\right)^{(k-1)/k} \tag{5-3f}$$

This is shown graphically in Figs. 5-4a and b. The figures show that flow availability is positive regardless of the temperature when $p/p_o \geq 1$; negative values can occur, but only for $p/p_o < 1$.

Using Eq. (5-3f) in Eq. (5-1d) gives the *reduced* availability

$$\frac{\bar{a}^*}{\bar{c}_p T_o} = F\left(\frac{T}{T_o}\right) + \left(\frac{k-1}{k}\right)\left[\ln\left(\frac{p}{p_o}\right) + \left(\frac{T}{T_o}\right)\left(\frac{p_o}{p} - 1\right)\right] \tag{5-3g}$$

This is shown graphically in Figs. 5-5a and b for the case $k = 1.4$. The figures show clearly that the availability cannot be negative in value.

Ideal Gas Mixtures. Consider a mixture of n ideal gases at temperature T and pressure p. The enthalpy and entropy per mole of mixture are, respectively,

$$\bar{h}^*(T) = \sum_{k=1}^{n} x_k \bar{h}_k^*(T)$$

$$\bar{s}^*(T, p) = \sum_{k=1}^{n} x_k \bar{s}_k^*(T, x_k p)$$

where x_k is the mole fraction of substance k in the mixture.

Inserting these relations into Eq. (5-1b)

$$\bar{a}_f^*(T, p) = \sum_{k=1}^{n} x_k[\bar{h}_k^*(T) - \bar{h}_k^*(T_o)] - T_o \sum_{k=1}^{n} x_k[\bar{s}_k^*(T, x_k p) - \bar{s}_k^*(T_o, x_k p_o)]$$

Then with

$$\bar{h}_k^*(T) - \bar{h}_k^*(T_o) = \int_{T_o}^{T} \bar{c}_{pk}(T) dT$$

and

$$\bar{s}_k^*(T, x_k p) - \bar{s}_k^*(T_o, x_k p_o) = \int_{T_o}^{T} \frac{\bar{c}_{pk}(T)}{T} dT - \bar{R} \ln(p/p_o)$$

it follows that

$$\bar{a}_f^* = \sum_{k=1}^{n} x_k \left[\int_{T_o}^{T} \left(1 - \frac{T_o}{T}\right) \bar{c}_{pk} dT\right] + \bar{R} T_o \ln(p/p_o)$$

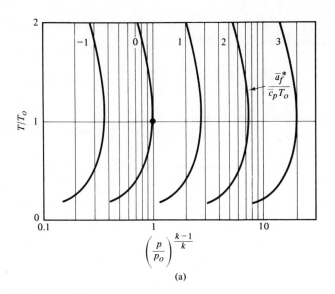

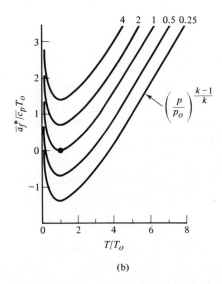

Figure 5-4 Reduced flow availability, $\bar{a}_f^*/\bar{c}_p T_o$, ideal gas (Eq. (5-3f)).

Upon identifying

$$\bar{a}_{fk}^*(T, p_o) = \int_{T_o}^{T} \left(1 - \frac{T_o}{T}\right) \bar{c}_{pk} dT$$

5-3 Ideal Gas Relations

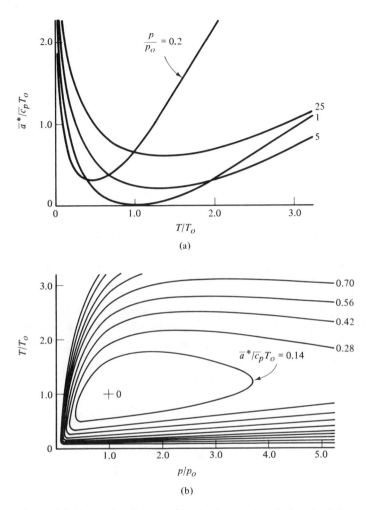

Figure 5-5 Reduced availability, $\bar{a}^*/\bar{c}_p T_o$, ideal gas (Eq. (5-3g), $k = 1.4$).

the last equation may be recast as

$$\bar{a}_f^* = \sum_{k=1}^{n} x_k \bar{a}_{fk}^*(T, p_o) + \bar{R} T_o \ln(p/p_o) \qquad (5\text{-}3h)$$

The quantities $\bar{a}_{fk}^*(T, p_o)$ are evaluated using the methods presented earlier in this section. See Problem 5-5 for further discussion.

5-4 Use of Generalized Property Charts

The objective of this section is to show how the flow availability for a real gas or vapor may be determined by use of generalized property charts.

The Generalized Charts. The variation of enthalpy with pressure at fixed temperature is

$$\left(\frac{\partial \bar{h}}{\partial p}\right)_T = \bar{v} - T\left(\frac{\partial \bar{v}}{\partial T}\right)_p$$

dh = Tds + v dp

$\left(\frac{\partial h}{\partial p}\right)_T = T\left(\frac{\partial s}{\partial p}\right)_T + v$

(see Problem 2-6). Integrating at fixed temperature

$= -T\left(\frac{\partial v}{\partial T}\right)_p + v$

$$\bar{h}(T, p) - \bar{h}(T, p') = \int_{p'}^{p} \left[\bar{v} - T\frac{\partial \bar{v}}{\partial T}\right]_p dp$$

The last equation may be rewritten on the left side as

$$[\bar{h}(T, p) - \bar{h}^*(T)] - [\bar{h}(T, p') - \bar{h}^*(T)] = \int_{p'}^{p} \left[\bar{v} - T\frac{\partial \bar{v}}{\partial T}\right]_p dp$$

where $\bar{h}^*(T)$ denotes the ideal gas model value of the enthalpy at the temperature under consideration. Then, making use of the idea that the properties of the real gas merge into those of the ideal gas model as pressure approaches zero

$$\lim_{p' \to 0} [\bar{h}(T, p') - \bar{h}^*(T)] = 0$$

it follows that the enthalpy of a real gas at a given state relative to that of its ideal gas model at the same state is given by

$$\bar{h}(T, p) - \bar{h}^*(T) = \int_{0}^{p} \left[\bar{v} - T\frac{\partial \bar{v}}{\partial T}\right]_p dp \quad (a)$$

To perform the integral appearing here requires p-v-T data in a form that $\partial \bar{v}/\partial T)_p$ can be accurately evaluated. Generalized compressibility data can be used for this purpose.

Using $Z = p\bar{v}/\bar{R}T$, Eq. (a) may be rewritten as the *enthalpy departure*

$$\frac{\bar{h}^*(T) - \bar{h}(T, p)}{T_c} = \bar{R} \int_{0}^{p_r} T_r^2 \left(\frac{\partial Z}{\partial T_r}\right)_{p_r} d(\ln p_r) \quad (b)$$

where $T_r = T/T_c$, $p_r = p/p_c$ (see Problem 5-6). The integral can be evaluated by numerical integration using generalized compressibility data, and is a function of the reduced temperature T_r and reduced pressure p_r. The result of this procedure is shown in Appendix Fig. B-2.

5-4 Use of Generalized Property Charts

Similar procedures can be used to evaluate entropy. Starting with the relation

$$\left(\frac{\partial \bar{s}}{\partial p}\right)_T = -\left(\frac{\partial \bar{v}}{\partial T}\right)_p$$

(see Problem 2-5), integration at fixed temperature gives

$$\bar{s}(T, p) - \bar{s}(T, p') = -\int_{p'}^{p} \left(\frac{\partial \bar{v}}{\partial T}\right)_p dp$$

This is valid for both the real gas and the corresponding ideal gas model.
For the ideal gas model, $\partial \bar{v}/\partial T)_p = \bar{R}/p$. So, for the same change in state, the ideal gas model entropy difference is

$$\bar{s}^*(T, p) - \bar{s}^*(T, p') = -\int_{p'}^{p} \frac{\bar{R}}{p} dp$$

Subtracting the last two expressions

$$[\bar{s}(T, p) - \bar{s}^*(T, p)] - [\bar{s}(T, p') - \bar{s}^*(T, p')] = \int_{p'}^{p} \left[\frac{\bar{R}}{p} - \left(\frac{\partial \bar{v}}{\partial T}\right)_p\right] dp$$

Using once again the idea that the properties of the real gas merge into those of the ideal gas model as pressure approaches zero

$$\lim_{p' \to 0} [\bar{s}(T, p') - \bar{s}^*(T, p')] = 0$$

it follows that the entropy of a real gas at a given state relative to that of its ideal gas model at the same state is given by

$$\bar{s}(T, p) - \bar{s}^*(T, p) = \int_{0}^{p} \left[\frac{\bar{R}}{p} - \left(\frac{\partial \bar{v}}{\partial T}\right)_p\right] dp$$

Using the same kind of procedures as for the case of enthalpy, the last equation may be rewritten as the *entropy departure*

$$\bar{s}^*(T, p) - \bar{s}(T, p) = \left[\frac{\bar{h}^*(T) - \bar{h}(T, p)}{T_r T_c}\right] + \bar{R} \int_{0}^{p_r} (Z - 1) d(\ln p_r)$$

The right side can be evaluated by numerical integration using generalized compressibility data, and is a function of reduced temperature T_r and reduced pressure p_r. The result of this procedure is shown in Appendix Fig. B-3.

Example 5-4. Using the generalized property charts, evaluate the change in enthalpy and in entropy per lbmol of nitrogen between a state defined by 520°R, 14.7 lbf/in.² and the state 300°R, 1000 lbf/in.². Compare with values obtained from Appendix Table B-4.

Solution. For nitrogen $T_c = 227.2°R$ and $p_c = 493 \text{ lbf/in.}^2$. Thus, at $300°R$, 1000 lbf/in.^2

$$p_{r2} = \frac{p_2}{p_c} = \frac{1000}{493} = 2.03$$

$$T_{r2} = \frac{T_2}{T_c} = \frac{300}{227.2} = 1.32$$

Using Appendix Figs. B-2 and B-3

$$\frac{\bar{h}^*(T_2) - \bar{h}(T_2, p_2)}{T_c} \approx 3 \frac{\text{Btu}}{(\text{lbmol})(°R)}$$

$$\bar{s}^*(T_2, p_2) - \bar{s}(T_2, p_2) \approx 1.7 \frac{\text{Btu}}{(\text{lbmol})(°R)}$$

At $520°R$, 14.7 lbf/in.^2,

$$p_{r1} = \frac{14.7}{493} = 0.03$$

$$T_{r1} = \frac{520}{227.2} = 2.29$$

Using Appendix Figs. B-2 and B-3

$$\frac{\bar{h}^*(T_1) - \bar{h}(T_1, p_1)}{T_c} \approx 0$$

$$\bar{s}^*(T_1, p_1) - \bar{s}(T_1, p_1) \approx 0$$

To determine the change in enthalpy, write

$$\bar{h}(T_2, p_2) - \bar{h}(T_1, p_1) = [\bar{h}^*(T_2) - \bar{h}^*(T_1)] - T_c \left[\left(\frac{\bar{h}^*(T_2) - \bar{h}(T_2, p_2)}{T_c} \right) \right.$$

$$\left. - \left(\frac{\bar{h}^*(T_1) - \bar{h}(T_1, p_1)}{T_c} \right) \right]$$

$$= \Delta\bar{h}^* - (227.2)(3)$$

$$= \Delta\bar{h}^* - 681.6 \text{ Btu/lbmol}$$

The ideal gas enthalpy change $\Delta\bar{h}^*$ is found from Appendix Table B-6B as $-1529.3 \text{ Btu/lbmol}$. So, $\Delta\bar{h} = -2210.9 \text{ Btu/lbmol}$. For comparison, with Appendix Table B-4, $\Delta\bar{h} = -2199.3 \text{ Btu/lbmol}$.

To determine the change in entropy, write

$$\bar{s}(T_2, p_2) - \bar{s}(T_1, p_1) = [\bar{s}^*(T_2, p_2) - \bar{s}^*(T_1, p_1)] - [(\bar{s}^*(T_2, p_2) - \bar{s}(T_2, p_2))$$

$$- (\bar{s}^*(T_1, p_1) - \bar{s}(T_1, p_1))]$$

$$= \Delta\bar{s}^* - 1.7$$

5-4 Use of Generalized Property Charts

where

$$\Delta \bar{s}^* = \bar{s}'(T_2) - \bar{s}'(T_1) - \bar{R} \ln (p_2/p_1)$$

From Appendix Table B-6B, $\bar{s}'(T_1) = 45.519$, $\bar{s}'(T_2) = 41.695$; so

$$\Delta \bar{s}^* = 41.695 - 45.519 - 1.986 \ln (1000/14.7) = -12.2 \frac{\text{Btu}}{(\text{lbmol})(°R)}$$

Finally, $\Delta \bar{s} = -13.9$ Btu/(lbmol)(°R). For comparison, with Appendix Table B-4, $\Delta \bar{s} = -13.87$ Btu/(lbmol)(°R).

As shown by these calculations, when used with care the generalized charts yield results in close agreement with the real gas property tables.

The Flow Availability Departure. The specific flow availability of a pure real gas at a given state, relative to the restricted dead state is

$$\bar{a}_f = \bar{h}(T, p) - \bar{h}(T_o, p_o) - T_o[\bar{s}(T, p) - \bar{s}(T_o, p_o)]$$

Using the ideal gas model at the *same state*

$$\bar{a}_f^* = \bar{h}^*(T) - \bar{h}^*(T_o) - T_o[\bar{s}^*(T, p) - \bar{s}^*(T_o, p_o)]$$

Subtracting these equations gives the departure of the flow availability from the ideal gas value

$$\bar{a}_f(T, p) = \bar{a}_f^*(T, p) - T_c \left\{ \left[\frac{\bar{h}^*(T) - \bar{h}(T, p)}{T_c} - \left(\frac{T_o}{T_c}\right)(\bar{s}^*(T, p) - \bar{s}(T, p)) \right] \right.$$
$$\left. - \left[\frac{\bar{h}^*(T_o) - \bar{h}(T_o, p_o)}{T_c} - \left(\frac{T_o}{T_c}\right)(\bar{s}^*(T_o, p_o) - \bar{s}(T_o, p_o)) \right] \right\} \quad (5\text{-}4a)$$

where T_c is the critical temperature. In this equation $[\bar{h}^*(T) - \bar{h}(T, p)]/T_c$ is recognized as the enthalpy departure and $[\bar{s}^*(T, p) - \bar{s}(T, p)]$ as the entropy departure. Accordingly, Appendix Figs. B-2 and B-3 can be used to evaluate the term in brackets; alternatively, the chart of Ref. [4] can be used.

Example 5-5. Using the generalized property charts, evaluate the flow availability for nitrogen at 300°R, 1000 lbf/in.². Let $T_o = 520°R$, $p_o = 14.7$ lbf/in.². Compare with the value obtained by use of property tables in Example 5-1: $a_f = 5012$ Btu/lbmol.

Solution. Using values from Example 5-4, Eq. (5-4a) gives

$$\bar{a}_f(T, p) = \bar{a}_f^*(T, p) - (227.2)\left\{\left[3 - \left(\frac{520}{227.2}\right)(1.7)\right] - 0\right\}$$
$$= \bar{a}_f^*(T, p) + 202.4$$

From Example 5-3, $\bar{a}_f^*(T,p) = 4817.2$ Btu/lbmol; accordingly, $\bar{a}_f(T,p) = 5019.6$ Btu/lbmol. In this case, a value is obtained well within 1% of that determined directly from the real gas tables.

As a final point it is noted that the compressibility factor Z of nitrogen at 300°R, 1000 lbf/in.² ($T_r = 1.32$, $p_r = 2.03$) is about 0.73. At this state use of the ideal gas model is not appropriate. Nonetheless, for the case at hand, the ideal gas approximation gives a_f to within 5%. The reason for this is that the enthalpy and entropy departures from ideality involved in evaluating the flow availability departure partially cancel.

The methods of this section can also be applied to gas mixtures by treating the overall mixture as if it were a pure substance having critical properties calculated by one of several alternative mixture rules. Kay's rule[4] is the simplest and most widely used of these, requiring only the calculation of a *pseudocritical temperature* T_c' and a *pseudocritical pressure* p_c',

$$T_c' = \sum x_k T_{ck}$$
$$p_c' = \sum x_k p_{ck}$$

where x_k is the mole fraction of substance k in the mixture and T_{ck}, p_{ck} are its critical temperature and pressure, respectively.

5-5 Closure

In this chapter various means for evaluating the thermomechanical availability and flow availability for pure substances and gas mixtures have been presented. Additional aspects of property evaluation directed to availability analysis, including those associated with reactive and nonreactive multicomponent systems, are discussed in Chaps. 6 and 7.

Selected References

1. KEENAN, J. H., "A Steam Chart for Second Law Analysis," *Mechanical Engineering*, **54**, March 1932, 195–204.
2. REISTAD, G. M., *Availability: Concepts and Applications*, doctoral dissertation, The University of Wisconsin-Madison, 1970, University Microfilms International, Ann Arbor, MI.
3. REISTAD, G. M., "A Property Diagram to Illustrate Irreversibilities in the R12 Refrigeration Cycle," *ASHRAE Transactions*, **78** (Part II), 1972, 97–101.

[4]For a discussion of mixture rules, including Kay's rule, see R. C. Reid and T. K. Sherwood, *The Properties of Gases and Liquids*, 2nd Ed., McGraw-Hill, New York, 1966, Chap. 7.

4. TAPIA, C. F., and M. J. MORAN, "A Generalized Property Chart to Evaluate Exergy," *Proc. 2nd World Congress of Chemical Engineering* (Montreal, Canada), October 1981.
5. THIRUMALESHWAR, M., "Exergy Method of Analysis and its Application to a Helium Cryorefrigerator," *Cryogenics*, **19**, 6, June 1979, 355–361.

Problems

5-1 Referring to Fig. 5-1, explain why the constant enthalpy lines are straight lines with the *same* negative slope.

5-2 Referring to Fig. 5-2, explain why the constant entropy lines are straight lines with the *same* positive slope.

5-3 Figure 5-3b shows that for nitrogen the saturated liquid flow availability at a specified temperature is greater in value than the flow availability at the corresponding saturated vapor state. Figure 5-3a shows the converse is true for water at states with $T > T_o$. Explain this difference.

5-4 Construct a detailed plot of (a) Fig. 5-4a, (b) Fig. 5-4b, (c) Fig. 5-5a, (d) Fig. 5-5b. The use of a computer is recommended.

5-5 Show that Eq. (5-3h) can be expressed as

$$\bar{a}_f^* = \bar{a}_f^*(T, p_o) + \bar{R}T_o \ln(p/p_o)$$

where

$$\bar{a}_f^*(T, p_o) = \int_{T_o}^{T} \left(1 - \frac{T_o}{T}\right) \bar{c}_{pm} dT$$

and $\bar{c}_{pm}(T)$ is the specific heat of the mixture (Eq. 2-4p).

5-6 Show that Eq. (a) (page 116) of Sec. 5-4 may be rewritten as Eq. (b).

5-7 For nitrogen at 250°R and 1000 lbf/in.^{2}, determine the flow availability in Btu/lb using Appendix Figs. B-2, 3. Compare with the value determined using Appendix Table B-4. Let $T_o = 60°F$ and $p_o = 14.7$ lbf/in.^{2}.

5-8 Determine the flow availability of ammonia in Btu/lb at 200°F and 30 lbf/in.^{2} relative to $T_o = 60°F$, $p_o = 15$ lbf/in.^{2}.

Chapter 6

Availability and Chemical Availability

The purpose of this chapter and the next is to extend the availability concept to account for the maximum work that can be done by a combined system of control mass and environment as the two come into *simultaneous* thermal, mechanical, and chemical equilibrium. As part of the presentation the *chemical* availability concept is introduced. A number of solved examples are included to illustrate the extended availability concept as well as its use in availability equations.

6-1 Introduction

The concept of availability introduced in Chap. 3 is limited in scope. It refers to the maximum work extractable from a combined system of control mass and environment as the control mass comes into thermal and mechanical equilibrium with the environment. The matter making up the control mass is not permitted to pass into or react chemically with the environment. To indicate this limitation, the designation *thermomechanical* availability is appropriate.

A more general formulation is realized in this chapter through the inclusion of the work that in principle can be developed as the contents of the control mass are permitted to pass into, but not react chemically with, the environment. A further extension of the availability concept is presented in Chap. 7. Consideration is given there to the possibility that one or more of the chemical substances present can enter into chemical reaction with environmental components.

In these extensions of the availability concept, the dead state is attained when the matter under consideration is in thermal, mechanical, *and* chemical equilibrium with the environment. The inclusion of chemical equilibrium necessitates a more detailed description of the state of the environment than given previously.

The illustrations and end of chapter problems of Chaps. 3 and 4 show that the thermomechanical availability is useful for engineering analysis. For example, it can be used to calculate irreversibilities and some kinds of second law efficiencies, and for evaluations involving the change in availability at fixed composition. Similarly, the illustrations and end of chapter problems of Chaps. 6, 7, and 8 bring out the usefulness of the extended availability concept. For instance, it is useful for the careful evaluation of availability losses in effluent streams, the calculation of all types of second law efficiencies, and the analysis of combustion and other chemical processes.

The presentations of this chapter and the next involve multicomponent simple compressible substances of variable composition. This topic is not normally treated in a first course in thermodynamics. Accordingly, before continuing the development of the availability concept some required background material is introduced in the next section.

6-2 Multicomponent Systems Background

This section presents background material on multicomponent simple compressible substances of variable composition in enough detail to support the discussions of this chapter. Additional background material aimed specifically at applications where composition varies due to chemical reaction is presented in Sec. 7-2.

Chemical Potential. Property relations for simple compressible systems of *fixed composition* are presented in Sec. 2-4. A key relationship is the first TdS equation

$$dU = TdS - pdV)_{\substack{\text{composition}\\\text{fixed}}}$$

This implies $U = U(S, V)$ and the relations

$$T = \frac{\partial U}{\partial S}\bigg)_{\substack{V,\text{composition}\\\text{fixed}}}, \qquad -p = \frac{\partial U}{\partial V}\bigg)_{\substack{S,\text{composition}\\\text{fixed}}} \qquad (a)$$

Under consideration now is a system that allows both energy and mass exchanges with its surroundings so that neither its size nor its chemical composition necessarily remains constant. The internal energy for this system is expressed as

$$U = U(S, V, N_1, N_2, \ldots, N_n)$$

where $N_k (k = 1, n)$ is the number of moles of substance k present.

The total differential of U is

$$dU = \left(\frac{\partial U}{\partial S}\right)_{V,N} dS + \left(\frac{\partial U}{\partial V}\right)_{S,N} dV + \sum_{k=1}^{n} \left(\frac{\partial U}{\partial N_k}\right)_{S,V,N_l} dN_k \qquad (b)$$

where the subscript N in the first two terms indicates that the composition is held fixed during differentiation, the subscript N_l in the third and higher terms means that all N's except N_k are held fixed.

Equation (b) involves $(2+n)$ partial derivatives. Upon comparison with Eqs. (a), the following identification is made for the first two

$$T = \left(\frac{\partial U}{\partial S}\right)_{V,N}, \qquad -p = \left(\frac{\partial U}{\partial V}\right)_{S,N} \qquad (c)$$

Turning to the remaining n derivatives of Eq. (b), the *chemical potential* of substance k is introduced as

$$\mu_k = \left(\frac{\partial U}{\partial N_k}\right)_{S,V,N_l} \qquad (k = 1, n) \qquad (d)$$

The chemical potential is a property because it is a function of other properties. Moreover, like the temperature and pressure it is an intensive property.

With Eqs. (c) and (d), Eq. (b) becomes

$$dU = TdS - pdV + \sum_{k=1}^{n} \mu_k dN_k \qquad (6\text{-}2a)$$

which is the first TdS equation extended to an open system consisting of n substances.

If each mole number is made α times as large, the *extensive* properties $V, S,$ and U all become α times larger. Thus, $U(\alpha S, \alpha V, \alpha N_1, \ldots, \alpha N_n) = \alpha U(S, V, N_1, \ldots, N_n)$, and the function U is said to be *homogeneous of degree one*. Differentiating with respect to α and using the chain rule on the left side

$$\frac{\partial U}{\partial(\alpha S)} S + \frac{\partial U}{\partial(\alpha V)} V + \sum_{k=1}^{n} \frac{\partial U}{\partial(\alpha N_k)} N_k = U$$

This holds for all values of α. In particular, it holds for $\alpha = 1$. Setting $\alpha = 1$ and using Eqs. (c) and (d)

$$U = TS - pV + \sum_{k=1}^{n} N_k \mu_k \qquad (6\text{-}2b)$$

The last result can be rearranged to

$$G = \sum_{k=1}^{n} N_k \mu_k \qquad (6\text{-}2c)$$

where $G = H - TS$ is the *Gibbs free energy*. For a single component system, Eq. (6-2c) reduces to

$$G/N = \mu \tag{e}$$

That is, for a pure component the chemical potential is equal to the Gibbs function per mole, $\mu = \bar{g}$.

Phase Equilibrium. This subsection develops the conditions that must be met for two or more phases to be in mutual stable equilibrium while open to the transfer of matter between them but with no other system. Central to the development is the notion that an isolated system attains equilibrium when its entropy reaches the maximum possible value consistent with the constraints imposed (Sec. 2-3).

Consider an isolated system consisting of two phases denoted by A and B. For the isolated system

$$dS = dS_A + dS_B \tag{f}$$

Let each phase consist of the same two substances, denoted 1 and 2. Then, the use of Eq. (6-2a) on the right side of Eq. (f) gives

$$dS = \left(\frac{dU_A}{T_A} + \frac{P_A}{T_A}dV_A - \frac{\mu_{1A}}{T_A}dN_{1A} - \frac{\mu_{2A}}{T_A}dN_{2A}\right)$$
$$+ \left(\frac{dU_B}{T_B} + \frac{P_B}{T_B}dV_B - \frac{\mu_{1B}}{T_B}dN_{1B} - \frac{\mu_{2B}}{T_B}dN_{2B}\right) \tag{g}$$

where μ_{1A} denotes the chemical potential of substance 1 in phase A and μ_{1B} is its chemical potential in phase B, and so on.

For the isolated system, the total energy, total volume, and total mass for each substance are constant; that is

$$dU_A + dU_B = 0$$
$$dV_A + dV_B = 0$$
$$dN_{1A} + dN_{1B} = 0$$
$$dN_{2A} + dN_{2B} = 0$$

With these, Eq. (g) can be written as

$$dS = \left(\frac{1}{T_A} - \frac{1}{T_B}\right)dU_A + \left(\frac{P_A}{T_A} - \frac{P_B}{T_B}\right)dV_A$$
$$- \left(\frac{\mu_{1A}}{T_A} - \frac{\mu_{1B}}{T_B}\right)dN_{1A} - \left(\frac{\mu_{2A}}{T_B} - \frac{\mu_{2B}}{T_B}\right)dN_{2A} \tag{h}$$

This shows that in seeking the maximum of S the independent variables are

U_A, V_A, N_{1A}, and N_{2A}. Necessary conditions for a maximum are

$$\frac{\partial S}{\partial U_A} = \frac{\partial S}{\partial V_A} = \frac{\partial S}{\partial N_{1A}} = \frac{\partial S}{\partial N_{2A}} = 0$$

The first of these conditions requires

$$\frac{1}{T_A} - \frac{1}{T_B} = 0$$

or $T_A = T_B$. From the second

$$\frac{p_A}{T_A} - \frac{p_B}{T_B} = 0$$

and since the temperatures are equal, $p_A = p_B$. With the remaining conditions

$$\frac{\mu_{1A}}{T_A} - \frac{\mu_{1B}}{T_B} = 0$$

$$\frac{\mu_{2A}}{T_A} - \frac{\mu_{2B}}{T_B} = 0$$

But since the temperatures are equal, a necessary condition for equilibrium is $\mu_{1A} = \mu_{1B}$ and $\mu_{2A} = \mu_{2B}$.

This reasoning may be extended to systems with more than two phases and with any number of components to give the necessary conditions for equilibrium.

Thermal equilibrium: The temperature of every phase is the same.

Mechanical equilibrium: The pressure of every phase is the same.

Chemical equilibrium: The chemical potential of each component has the same value in every phase.

Consider again the isolated system consisting of phases A and B, and let the two phases be in thermal and mechanical equilibrium. Also, let the chemical potential of substance 2 be the same in both phases. The entropy change, Eq. (h), is then simply

$$dS = \frac{1}{T}(\mu_{1B} - \mu_{1A})dN_{1A}$$

where dN_{1A} represents the amount of substance 1 transferred from phase B to phase A. Entropy increases in a spontaneous process of the isolated system: $dS > 0$. Accordingly, when $\mu_{1B} > \mu_{1A}$, $dN_{1A} > 0$; substance 1 passes from phase B to phase A. When $\mu_{1B} < \mu_{1A}$, $dN_{1A} < 0$; substance 1 passes from phase A to phase B. From this it can be seen that the chemical potential is a measure

6-2 Multicomponent Systems Background

of the *escaping tendency* of a substance. Any substance will try to move from the phase having the higher chemical potential for that substance to the phase having lower chemical potential.

Ideal Gas Mixtures. Selecting as independent variables the temperature T, pressure p, and the number of moles of each substance N_k, the extensive properties U, H, and S for an ideal gas mixture are, respectively

$$U = \sum_{k=1}^{n} N_k \bar{u}_k(T)$$

$$H = \sum_{k=1}^{n} N_k \bar{h}_k(T)$$

$$S = \sum_{k=1}^{n} N_k \bar{s}_k(T, p_k)$$

where $p_k = x_k p$ is the partial pressure of substance k in the mixture.

Inserting the ideal gas expressions for H and S into the Gibbs function $G = H - TS$ results in

$$\begin{aligned} G &= \sum_{k=1}^{n} N_k \bar{h}_k(T) - T \sum_{k=1}^{n} N_k \bar{s}_k(T, p_k) \\ &= \sum_{k=1}^{n} N_k \bar{g}_k(T, p_k) \end{aligned} \quad (6\text{-}2\text{d})$$

where $\bar{g}_k(T, p_k) = \bar{h}_k(T) - T\bar{s}_k(T, p_k)$.

With Eq. (6-2c) it follows that

$$\mu_k = \bar{g}_k(T, p_k) \quad (6\text{-}2\text{e})$$

That is, the chemical potential of component k in an ideal gas mixture is equal to its Gibbs function per mole of k, evaluated at the mixture temperature and the partial pressure p_k of k in the mixture.

The chemical potential of component k also can be evaluated via Eq. (6-2f) which is obtained as follows. Integrating the third equation of Eqs. (2-4d) at fixed temperature from pressure p' to pressure p_k $(= x_k p)$ results in

$$\bar{g}_k(T, p_k) - \bar{g}_k(T, p') = \int_{p'}^{p_k} \bar{v}\, dp$$

The integral is readily performed for an ideal gas since $\bar{v} = \bar{R}T/p$; thus

$$\bar{g}_k(T, p_k) = \bar{g}_k(T, p') + \bar{R}T \ln(p_k/p')$$

With this, Eq. (6-2e) gives

$$\begin{aligned} \mu_k &= \bar{g}_k(T, p') + \bar{R}T \ln(p_k/p') \\ &= \bar{g}_k(T, p') + \bar{R}T \ln(x_k p/p') \end{aligned} \quad (6\text{-}2\text{f})$$

Partial Molal Properties. The simple ideal gas relations just considered are not valid for mixtures that cannot be modeled as ideal gas mixtures. In this subsection property relations of more general applicability are presented.

Selecting as independent variables the temperature T, pressure p, and the number of moles of each substance N_k, the extensive property Y can be expressed as, $Y(T, p, N_1, N_2, \ldots, N_n)$. By definition, the *partial molal property* $\bar{Y}_k$ of component k is

$$\bar{Y}_k = \left.\frac{\partial Y}{\partial N_k}\right)_{T, p, N_l}$$

Since Y is homogeneous of degree one in the N's: $\alpha Y(T, p, N_1, \ldots, N_n) = Y(T, p, \alpha N_1, \ldots, \alpha N_n)$, it can be written in terms of partial molal properties

$$Y(T, p, N_1, \ldots, N_n) = \sum_{k=1}^{n} N_k \bar{Y}_k(T, p, N_1, \ldots, N_n) \tag{i}$$

(Problem 6-1).

Selecting Y in turn to be enthalpy, entropy, and the Gibbs function,

$$H(T, p, N_1, \ldots, N_n) = \sum_{k=1}^{n} N_k \bar{H}_k(T, p, N_1, \ldots, N_n)$$

$$S(T, p, N_1, \ldots, N_n) = \sum_{k=1}^{n} N_k \bar{S}_k(T, p, N_1, \ldots, N_n) \tag{6-2g}$$

$$G(T, p, N_1, \ldots, N_n) = \sum_{k=1}^{n} N_k \bar{G}_k(T, p, N_1, \ldots, N_n)$$

With Eq. (6-2c) it follows that

$$\begin{aligned}\mu_k &= \bar{G}_k \\ &= \bar{H}_k - T\bar{S}_k\end{aligned} \tag{6-2h}$$

That is, the chemical potential of component k in a mixture is equal to its partial molal Gibbs function.

Semipermeable Membranes. There are in existence a variety of inorganic and organic membranes, called *semipermeable membranes*, which are known to allow the passage of certain substances, but prevent others from doing so. For instance, one sea water desalination method uses membranes permeable to positive ions and others permeable to negative ions. There are many other industrial applications using semipermeable membranes. These membranes also exist in the human body and in other living things. When referred to in this book a semipermeable membrane is assumed to be perfectly effective in the sense that only one substance can pass through.

6-3 Availability

As an application, consider a multicomponent system at temperature T and pressure p separated from a pure substance k, also at temperature T, by a membrane permeable only to k. For chemical equilibrium the chemical potential of pure k is equal to its chemical potential within the mixture

$$\mu_{k,\text{pure}} = \mu_{k,\text{mixture}}$$

For the special case of ideal gases, use of this relation along with Eq. (6-2e) leads to the conclusion that the pressure of the pure phase equals the partial pressure of substance k in the mixture (see Problem 6-2). These results are used in the next section in the discussions centering on Problem 6-4 and Fig. 6-2.

6-3 Availability

Introduction. The maximum work that can be extracted from a combined system of control mass and environment as the control mass passes from a given state to the restricted dead state

$$= (E - U_o) + p_o(V - V_o) - T_o(S - S_o)$$

This is the thermomechanical availability.

At the restricted dead state the control mass is in thermal and mechanical equilibrium with the environment, but not necessarily in chemical equilibrium with it. In principle, a difference between the composition of the control mass at the restricted dead state and that of the environment can be exploited to obtain additional work. The maximum work obtainable in this way is the *chemical* availability.

The concept of availability is extended in this section. The goal is to evaluate the maximum work that can be extracted from the combined system as the control mass is brought from some given state into complete equilibrium with the environment, which for now is assumed to contain *at least* those substances present within the control mass.

The Environment. When the thermomechanical availability is the objective, the composition of the environment is of no interest. Its state is adequately defined by temperature T_o and pressure p_o. But in evaluations where the control mass is brought into thermal, mechanical, *and* chemical equilibrium with the environment, it is also necessary to consider the composition of both the control mass and the environment. For such applications, the environment is *modeled* as follows.

The environment is large in extent. All parts are at rest relative to one another. It experiences only internally reversible processes in which its intensive state remains unaltered. The environment is made up entirely of common substances which exist in abundance within the atmosphere, the oceans, and the crust of the Earth. The substances are in their stable forms as they exist naturally.

The state of the environment is described by the extensive properties internal energy U^o, volume V^o, and entropy S^o and by the intensive properties temperature T_o, pressure p_o, and chemical potentials μ_k^o ($k = 1, n$) of the n substances present.[1] Using Eq. (6-2b), these properties are related as follows

$$U^o = T_o S^o - p_o V^o + \sum_{k=1}^{n} N_k^o \mu_k^o \tag{a}$$

where N_k^o is the number of moles of substance k.

Derivation. Figure 6-1 shows a combined system consisting of a control mass and an environment. Heat and work interactions can take place between the control mass and environment, but only work interactions are permitted across the control surface of the combined system. The state of the control mass is described by the extensive properties energy E, volume V, and entropy S and by the intensive properties temperature T, pressure p, and chemical potentials μ_k ($k = 1, n$) of the n substances present. To keep the presentation simple it is assumed the environment contains *exactly* those substances in the control mass, but the main results are also valid when the environment contains substances in addition to those of the control mass.

Let the control mass interact with the environment until both the control

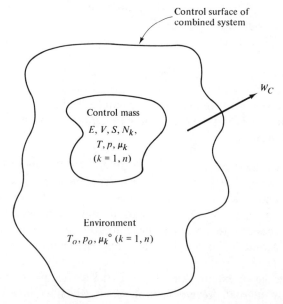

Figure 6-1 Combined system.

[1] Except for temperature and pressure, superscript ^{o} denotes properties of the environment. Subscript $_o$ denotes properties of the control mass at the restricted dead state.

6-3 Availability

mass and environment come to equilibrium. The work derived from the combined system is determined from an energy equation as

$$W_c = -\Delta E_c \tag{b}$$

where ΔE_c is the energy change of the combined system.

Initially, the entropy of the combined system is $S^o + S$, the volume is $V^o + V$, and the energy is $U^o + E$, where U^o is given by Eq. (a). At equilibrium the energy of the combined system is

$$\acute{U}^o = T_o \acute{S}^o - p_o \acute{V}^o + \sum_{k=1}^{n} \acute{N}_k^o \mu_k^o \tag{c}$$

where

$$\begin{aligned} \acute{S}^o &= S^o + S + \sigma_c \\ \acute{V}^o &= V^o + V \\ \acute{N}_k^o &= N_k^o + N_k \quad (k = 1, n) \end{aligned} \tag{d}$$

The term σ_c in the first of Eqs. (d) accounts for the entropy produced due to irreversibilities. The second equation follows because there is no change in total volume of the combined system, and the third because the amount of each substance present is conserved.

Collecting Eqs. (a) to (d)

$$W_c = \left(E + p_o V - T_o S - \sum_{k=1}^{n} N_k \mu_k^o \right) - T_o \sigma_c \tag{e}$$

(Problem 6-3).

Since $\sigma_c \geq 0$, it follows that

$$W_c \leq E + p_o V - T_o S - \sum_{k=1}^{n} N_k \mu_k^o$$

with the maximum work of the combined system being

$$(W_c)_{\max} = E + p_o V - T_o S - \sum_{k=1}^{n} N_k \mu_k^o$$

The maximum is obtained only when the process of the combined system is in every respect internally reversible. By definition, the availability equals $(W_c)_{\max}$. That is

$$A = E + p_o V - T_o S - \sum_{k=1}^{n} N_k \mu_k^o \tag{6-3a}$$

Discussion. The term *availability* is used in this book for the expression on the right side of Eq. (6-3a). However, it has been referred to in the literature by other names; among these are *essergy* (essence of energy) and *exergy*.

Availability is the amount of work obtainable from the combined system when the control mass is brought into complete equilibrium with the environment by means of reversible processes involving interaction only with the environment. Its magnitude depends on two states, the state of the control mass and that of the environment, and is a measure of the departure of the state of the control mass from that of the environment. The magnitude is independent of the particular series of ideal processes followed as equilibrium is established.

Chemical Availability. For some applications it is convenient to consider the availability as the sum of two contributions, the *thermomechanical* availability and the *chemical* availability. To develop this idea, let the control mass be brought by means of ideal processes from its initial state to a condition of thermal and mechanical equilibrium with the environment—that is, to the restricted dead state. As detailed in Sec. 3-4, the work obtained in this step is the thermomechanical availability, denoted for clarity here as A^{tm}

$$A^{tm} = (E - U_o) + p_o(V - V_o) - T_o(S - S_o) \tag{6-3b}$$

Using Eq. (6-2b) this can be written as

$$A^{tm} = E + p_o V - T_o S - \sum_{k=1}^{n} N_k \mu_{ko} \tag{f}$$

where μ_{ko} is the chemical potential of substance k within the mixture at the restricted dead state.

Next, let the control mass be brought into chemical equilibrium with the environment. The maximum work obtainable in this step is the chemical availability, A^{ch}. Subtracting Eq. (f) from Eq. (6-3a), the *chemical availability* is

$$A^{ch} = \sum_{k=1}^{n} N_k (\mu_{ko} - \mu_k^o) \tag{6-3c}$$

Another way to look at the chemical availability is that it is the *minimum work input* required to extract the mixture at the restricted dead state from the substances within the environment (see Problem 6-4). This is illustrated in the following subsection for the ideal gas case.

Ideal Gases. When substance k exists as a member of an ideal gas mixture *both* at the restricted dead state and in the environment, a particularly simple expression is obtained for the chemical availability.

With Eq. (6-2f)

$$\mu_{ko} = \bar{g}_k(T_o, p_o) + \bar{R} T_o \ln (p_k/p_o)$$
$$\mu_k^o = \bar{g}_k(T_o, p_o) + \bar{R} T_o \ln (p_k^o/p_o)$$

where p_k is the partial pressure of substance k in the mixture at the restricted

6-3 Availability

dead state and p_k^o is the partial pressure in the environment. Substituting into Eq. (6-3c)

$$A^{ch} = \bar{R}T_o \sum_{k=1}^{n} N_k \ln(p_k/p_k^o)$$

Or, with $p_k = x_k p_o$, $p_k^o = x_k^o p_o$

$$A^{ch} = \bar{R}T_o \sum_{k=1}^{n} N_k \ln(x_k/x_k^o) \tag{6-3d}$$

The following development provides a physical picture for the determination of this result. Let the objective be to determine the *minimum work input* required at steady-state to extract a specified ideal gas mixture from the environment and bring it to the restricted dead state. Figure 6-2 shows an extraction unit for this purpose comprising n subunits, one for each substance, of which k is typical. For minimum work input each subunit necessarily operates reversibly. Central to its operation is a membrane permeable only to substance k (Sec. 6-2).

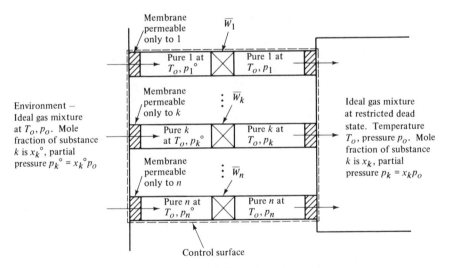

Figure 6-2 Extraction unit.

In subunit k ($k = 1, n$) work is required to pass substance k from a pressure p_k^o to a pressure p_k. For reversible operation, the work *input per mole* of substance k is

$$\bar{W}_k = \int_{p_k^o}^{p_k} \bar{v} \, dp \tag{g}$$

a result which is readily developed by use of the steady-state forms of the energy and entropy equations (see Problem 6-5). Since $\bar{v} = \bar{R}T_o/p$, it follows that

$$\bar{W}_k = \bar{R}T_o \ln(p_k/p_k^o)$$
$$= \bar{R}T_o \ln(x_k/x_k^o)$$

For N_k moles

$$W_k = \bar{R}T_o[N_k \ln(x_k/x_k^o)]$$

Finally, summing over the n subunits gives Eq. (6-3d).

6-4 Control Volume Analyses

This section and the next continue the development of the availability concept with emphasis on the flow availability and its use in control volume analyses.

Flow Availability. Using Eq. (6-3a), the availability expressed on a per mole of mixture basis is

$$\bar{a} = \bar{u} + p_o\bar{v} - T_o\bar{s} - \sum_{k=1}^{n} x_k \mu_k^o \tag{6-4a}$$

where x_k is the mole fraction of substance k in the mixture. The kinetic and potential energy terms are not shown.

As discussed in Sec. 3-6, a stream of matter passing through a portion of a control surface has associated with it the availability of the flowing mass and a contribution due to the work interaction there given by $(p - p_o)\bar{v}$. The flow availability is the sum of the two

$$\bar{a}_f = \bar{a} + (p - p_o)\bar{v} \tag{a}$$

Introducing Eq. (6-4a), this can be rewritten as

$$\bar{a}_f = \bar{h} - T_o\bar{s} - \sum_{k=1}^{n} x_k \mu_k^o \tag{6-4b}$$

Alternatively, using Eqs. (6-3b and c) the availability expressed on a per mole of mixture basis is

$$\bar{a} = [(\bar{u} - \bar{u}_o) + p_o(\bar{v} - \bar{v}_o) - T_o(\bar{s} - \bar{s}_o)] + \sum_{k=1}^{n} x_k(\mu_{ko} - \mu_k^o)$$

Introducing this into Eq. (a) and reducing

$$\bar{a}_f = (\bar{h} - \bar{h}_o) - T_o(\bar{s} - \bar{s}_o) + \sum_{k=1}^{n} x_k(\mu_{ko} - \mu_k^o) \tag{6-4c}$$

This shows that the flow availability can be viewed as the sum of the thermomechanical flow availability and the chemical availability.

6-4 Control Volume Analyses

Special Cases. Three types of inlet and exit streams frequently encountered in engineering applications are streams of water vapor, streams of liquid water, and streams consisting of an ideal gas mixture. In this subsection and the next, equations are developed for the evaluation of the flow availability for these cases.

For purposes of illustration, the environment is modeled as shown in Table 6-1. It is assumed to consist of a gas phase including water vapor and a number of dry components. The gas phase forms an ideal gas mixture.

Table 6-1

Environment: Ideal Gas Mixture at $T_o = 298.15$ K (77°F), $p_o = 1$ atm

	Substance	Mole Fraction
	$H_2O(v)$	x_v^o
Dry components	N_2	$x_{N_2}^o$
	O_2	$x_{O_2}^o$
	CO_2	$x_{CO_2}^o$
	$\vdots$	$\vdots$

For a stream of water vapor or liquid water at temperature T and pressure p, Eq. (6-4b) written on a *unit mass* basis is

$$a_f = h(T, p) - T_o s(T, p) - \mu_v^o$$

where μ_v^o is the chemical potential of water in the environment. The kinetic and potential energy terms are not shown.

Since the environment is modeled as an ideal gas mixture, the chemical potential μ_v^o appearing in the last equation can be evaluated by use of Eq. (6-2f)

$$\mu_v^o = g_g(T_o) + R_v T_o \ln\left(\frac{x_v^o p_o}{p_g(T_o)}\right)$$

where $p_g(T_o)$ is the saturation pressure of water at temperature T_o, $R_v = \bar{R}/M_v$, x_v^o is the mole fraction of the water vapor in the environment, and g_g is the Gibbs function

$$g_g(T_o) = h_g(T_o) - T_o s_g(T_o)$$

with the subscript g denoting saturated vapor property values.

Collecting the last three equations

$$a_f = [h(T, p) - h_g(T_o)] - T_o[s(T, p) - s_g(T_o)] - R_v T_o \ln\left(\frac{x_v^o p_o}{p_g(T_o)}\right) \quad (6\text{-}4d)$$

This is applicable to a stream of water vapor as well as to a stream of liquid water. In either case, the enthalpy and entropy values can be evaluated by means of Appendix Table B-2.

Idealizing the *liquid only* stream as being incompressible, Eq. (6-4d) can be rewritten with the help of

$$h(T, p) = h_f(T) + v_f[p - p_g(T)]$$
$$s(T, p) = s_f(T) \qquad (2\text{-}4s)$$

to give

$$a_f)_{\substack{\text{liquid}\\\text{water}}} = [h_f(T) - h_g(T_o)] + v_f[p - p_g(T)]$$
$$- T_o[s_f(T) - s_g(T_o)] - R_v T_o \ln\left(\frac{x_v^o p_o}{p_g(T_o)}\right) \qquad (6\text{-}4e)$$

where, with the exception of a_f, f denotes saturated liquid property values at temperature T.

Example 6-1. Referring to Example 3-8, determine the flow availability at the valve inlet and the valve exit relative to the environment of Table 6-1 with $x_v^o = 0.0125$. Discuss.

Solution. The required results can be obtained with Eq. (6-4d), but to relate the current calculations more closely to those of Example 3-8 it is first rewritten with[2]

$$h_f(T_o) - T_o s_f(T_o) = h_g(T_o) - T_o s_g(T_o)$$

as

$$a_f = \{[h(T, p) - h_f(T_o)] - T_o[s(T, p) - s_f(T_o)]\} - R_v T_o \ln(x_v^o p_o/p_g(T_o)) \quad (b)$$

The underlined term in Eq. (b) is evaluated in Example 3-8 at both the inlet and the exit of the valve; the values are, respectively, 431.2 Btu/lb and 330.9 Btu/lb. The flow availability at the valve inlet is then

$$a_{f1} = 431.2 - R_v T_o \ln(x_v^o p_o/p_g(T_o))$$

With $x_v^o = 0.0125$ and $p_g(T_o) = 0.46$ lbf/in.^{2},

$$a_{f1} = 431.2 - \left(\frac{1.986}{18}\right)(537) \ln(0.4)$$
$$= 431.2 + 54.3$$
$$= 485.5 \text{ Btu/lb}$$

---

[2] For the liquid and vapor phases of a pure substance in equilibrium at fixed temperature and pressure, $g_f(T) = g_g(T)$.

6-4 Control Volume Analyses

Similarly, $a_{f2} = 385.2$ Btu/lb. Thus, the two flow availabilities are each over 12% greater than calculated in Example 3-8 relative to the restricted dead state. The importance of this is brought out in the following discussion.

Discussion. When computing a second law efficiency of the form $\epsilon' = a_{f2}/a_{f1}$ (Sec. 4-4), the use of availabilities determined relative to the restricted dead state can be somewhat misleading. Thus, using the values of this example, $\epsilon' = 79\%$, whereas with those of Example 3-8, $\epsilon' = 77\%$.

Moreover, if the steam were to be subsequently vented to the surroundings, the choice of the dead state is important in evaluating the fraction of the availability input irrevocably lost in the effluent stream. Calculations based on the restricted dead state can result in significant misevaluations.

On the other hand, except for T_o the value of the irreversibility for the valve is independent of the dead state. This may be verified by using Eq. (b) to evaluate the right side of

$$\dot{I}/\dot{m}_f = a_{f1} - a_{f2}$$

Noting that $h_2 = h_1$, and that the term $\{h_f(T_o) - T_o s_f(T_o) - R_v T_o \ln(x_v^o p_o / p_g(T_o))\}$ cancels, it follows that $\dot{I}/\dot{m}_f = T_o(s_2 - s_1)$.

Ideal Gas Mixture. Consider next a stream which forms an ideal gas mixture at temperature T and pressure p. The products of combustion of hydrocarbon fuels provide an example of this kind of stream.

Applying Eq. (6-4b), the flow availability *per mole of mixture* is

$$\bar{a}_f = \sum_{k=1}^{n} x_k[\bar{h}_k(T) - T_o \bar{s}_k(T, x_k p) - \mu_k^o]$$

where x_k is the mole fraction of substance k in the gas mixture. Upon evaluating the chemical potentials with Eq. (6-2e) this becomes

$$\bar{a}_f = \sum_{k=1}^{n} x_k\{\bar{h}_k(T) - \bar{h}_k(T_o) - T_o[\bar{s}_k(T, x_k p) - \bar{s}_k(T_o, x_k^o p_o)]\}$$

Finally, using Eq. (2-4l) to evaluate the entropy differences

$$\bar{a}_f = \sum_{k=1}^{n} x_k\left\{\bar{h}_k(T) - \bar{h}_k(T_o) - T_o[\bar{s}_k'(T) - \bar{s}_k'(T_o)] + \bar{R}T_o \ln\left(\frac{x_k p}{x_k^o p_o}\right)\right\}$$

$$= \sum_{k=1}^{n} x_k\left\{\underline{\bar{h}_k(T) - \bar{h}_k(T_o) - T_o[\bar{s}_k'(T) - \bar{s}_k'(T_o)]} + \underline{\bar{R}T_o \ln\left(\frac{x_k}{x_k^o}\right)}\right\} + \underline{\bar{R}T_o \ln\left(\frac{p}{p_o}\right)}$$

(6-4f)

The first underlined term on the right side accounts for the stream temperature being different than T_o, it can be evaluated with the ideal gas tables. The second underlined term accounts for the stream composition being different from that of the environment, and the third for $p \neq p_o$.

***

Example 6-2. Combustion products exit a reactor at temperature T and a pressure of 1 atm with a composition *per mole of fuel* consumed described by (1CO_2, 2$H_2O(v)$, 10.53N_2, 0.8O_2). Evaluate the flow availability relative to an environment consisting of an ideal gas mixture of water vapor and dry air at 77°F, 1 atm described as follows: $x_v^o = 0.0303$, $x_{N_2}^o = 0.7567$, $x_{O_2}^o = 0.2035$, $x_{CO_2}^o = 0.0003$.

Solution. The required result is obtained using Eq. (6-4f). There are two separate contributions, one depending on temperature T and the other on the composition of the combustion products. The contribution associated with the composition, per mole of fuel consumed

$$= \{\bar{R}T_o[\ln(x_{CO_2}/x_{CO_2}^o) + 2\ln(x_v/x_v^o) + 10.53\ln(x_{N_2}/x_{N_2}^o) + 0.8\ln(x_{O_2}/x_{O_2}^o)]\}$$

From the given composition of the products, $x_{CO_2} = 0.0698$, $x_v = 0.1396$, $x_{N_2} = 0.7348$, $x_{O_2} = 0.0558$. Substituting values, this contribution is 7636 Btu/lbmol of fuel.

The contribution depending on temperature T

$$= \{[\bar{h}(T) - \bar{h}(T_o) - T_o(\bar{s}'(T) - \bar{s}'(T_o))]_{CO_2} +$$
$$2[\bar{h}(T) - \bar{h}(T_o) - T_o(\bar{s}'(T) - \bar{s}'(T_o))]_v +$$
$$10.53[\bar{h}(T) - \bar{h}(T_o) - T_o(\bar{s}'(T) - \bar{s}'(T_o))]_{N_2} +$$
$$0.8[\bar{h}(T) - \bar{h}(T_o) - T_o(\bar{s}'(T) - \bar{s}'(T_o))]_{O_2}\}$$

Using Appendix Table B-6 to evaluate properties, the temperature dependent contribution

$$= \{[(\bar{h}(T) - 4027.5) - 537(\bar{s}'(T) - 51.032)]_{CO_2} +$$
$$2[(\bar{h}(T) - 4258) - 537(\bar{s}'(T) - 45.079)]_v +$$
$$10.53[(\bar{h}(T) - 3729.5) - 537(\bar{s}'(T) - 45.743)]_{N_2} +$$
$$0.8[(\bar{h}(T) - 3725.1) - 537(\bar{s}'(T) - 48.982)]_{O_2}\}$$

Letting temperature T vary, the results given in the table are obtained.

T(°R)	740	865	1400	2100	2820	3371
Temperature dependent contribution to the flow availability (Btu/lbmol of fuel consumed)	3243	7622	38,521	99,542	169,319	230,276

6-5 Psychrometric Processes

Up to about 865°R it is the contribution associated with composition which is more important, above 865°R the contribution associated with temperature is dominant, becoming more so as temperature increases.

When a stream of matter having a nonzero value for availability is vented directly to the surroundings an irrevocable loss of some portion of the availability input to the device or plant results. For instance, a fraction of the availability entering a power plant with the fuel exits in the stack gases. Example 6-2 emphasizes that both composition and temperature are important in evaluating such an effluent stream. A full availability analysis for a power plant in which the calculations of Example 6-2 play a part is given in Sec. 7-4.

6-5 Psychrometric Processes

Introduction. The objective of this section is to adapt the concepts developed in Secs. 6-3 and 6-4 to psychrometric processes, that is to processes which involve a binary mixture of water vapor and dry air. Though dry air itself is a mixture, for simplicity it is treated in this section as if it was a pure substance.

For a wide range of practical psychrometric processes, particularly those involving heating, ventilating, and air conditioning (HVAC), moist air can be regarded as an ideal gas mixture obeying the Dalton mixture model (Sec. 2-4). The composition of the mixture is described by the mole fractions x_v and x_a of water vapor and dry air, respectively. The composition is also described by the *humidity ratio* ω

$$\omega = \frac{m_v}{m_a} \qquad\qquad (a)$$

$$= \frac{M_v x_v}{M_a x_a} = 0.622 \frac{x_v}{x_a}$$

where m_v is the mass of water vapor, m_a is the mass of dry air, M_v is the molecular weight of water ($= 18.016$), and M_a is the molecular weight of dry air ($= 28.967$).

Using $x_a + x_v = 1$, Eq. (a) can be solved to give

$$x_v = \frac{\tilde{\omega}}{1 + \tilde{\omega}}$$

$$x_a = \frac{1}{1 + \tilde{\omega}} \qquad\qquad (b)$$

where $\tilde{\omega} = \omega M_a / M_v = 1.6078\omega$.

The *relative humidity* ϕ is evaluated as the ratio of the partial pressure p_v of the vapor in the mixture to the saturation pressure p_g of the vapor at the temperature of the mixture

$$\phi = \frac{p_v}{p_g(T)} \qquad (c)$$

Environment. For purposes of illustration, the environment is modeled in this section as follows. It is assumed to consist of an ideal gas mixture of water vapor and dry air at temperature T_o and pressure p_o. The composition is defined by the water vapor mole fraction x_v^o, or equivalently by the humidity ratio ω^o or relative humidity ϕ^o.

Flow Availability for Moist Air. In this subsection equations are developed for the evaluation of the flow availability for moist air streams. It is convenient to begin with Eq. (6-4f) written for an ideal gas mixture of dry air and water vapor

$$\begin{aligned}\bar{a}_f = & \, x_a\{[\bar{h}_a(T) - \bar{h}_a(T_o)] - T_o[\bar{s}'_a(T) - \bar{s}'_a(T_o)]\} \\ & + \bar{R}T_o \ln(x_a/x_a^o) + x_v\{[\bar{h}_v(T) - \bar{h}_v(T_o)] - T_o[\bar{s}'_v(T) - \bar{s}'_v(T_o)]\} \qquad (d) \\ & + \bar{R}T_o \ln(x_v/x_v^o) + \bar{R}T_o \ln(p/p_o)\end{aligned}$$

Assuming the specific heat of air, $\bar{c}_{pa}$, is constant

$$\bar{h}_a(T) - \bar{h}_a(T_o) = \bar{c}_{pa}(T - T_o)$$
$$\bar{s}'_a(T) - \bar{s}'_a(T_o) = \bar{c}_{pa} \ln(T/T_o)$$

Similar expressions are obtained for the water vapor for constant $\bar{c}_{pv}$. Inserting these into Eq. (d) and rearranging terms

$$\begin{aligned}\bar{a}_f = & \, T_o\{(x_a\bar{c}_{pa} + x_v\bar{c}_{pv})[(T/T_o) - 1 - \ln(T/T_o)] + \bar{R}\ln(p/p_o)\} \\ & + \bar{R}T_o[x_a \ln(x_a/x_a^o) + x_v \ln(x_v/x_v^o)] \qquad (6\text{-}5a)\end{aligned}$$

Repeating the development, the flow availability on a *per unit mass of dry air* basis is

$$\begin{aligned}a_f = & \, T_o\left\{(c_{pa} + \omega c_{pv})\left[\frac{T}{T_o} - 1 - \ln\left(\frac{T}{T_o}\right)\right] + (1 + \tilde{\omega})R_a \ln(p/p_o)\right\} \\ & + R_a T_o\left\{(1 + \tilde{\omega}) \ln\left(\frac{1 + \tilde{\omega}^o}{1 + \tilde{\omega}}\right) + \tilde{\omega} \ln\left(\frac{\tilde{\omega}}{\tilde{\omega}^o}\right)\right\} \qquad (6\text{-}5b)\end{aligned}$$

(Problem 6-6). In obtaining this, Eqs. (b) have been used to eliminate the mole fractions in favor of the parameter $\tilde{\omega}$ (= 1.6078ω). In Eq. (6-5b), $R_a = \bar{R}/M_a$, and c_{pa} and c_{pv} denote, respectively, the specific heat of dry air and water vapor on a unit mass basis.

Equation (6-5b) has been used to prepare Fig. 6-3 which is a psychrometric chart with contours of flow availability per unit mass of dry air at one atmos-

6-5 Psychrometric Processes

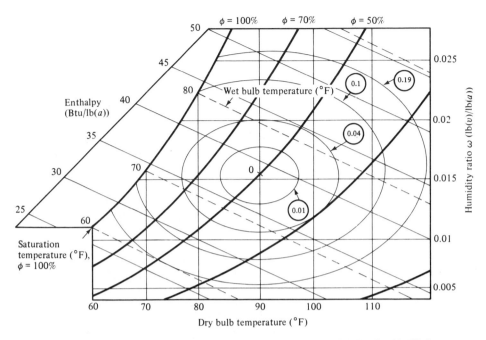

Figure 6-3 Standard psychrometric chart with contours of flow availability, Eq. (6-5b), in Btu/lb(a) at $p = 1$ atm ($T_o = 90°F$, $p_o = 1$ atm, $\omega^o = 0.0153$).

phere superimposed. The values are relative to an environment defined by $T_o = 90°F$, $p_o = 1$ atm, $\omega^o = 0.0153$ lb(v)/lb(a) (representative of daytime summer conditions in some parts of the U.S.).[3]

Application. In the example to follow the principles developed in this section are illustrated by application to a system present in many industrial settings, a cooling tower. The solution also shows the use of the extended availability concept in an availability equation.

Example 6-3. Determine a_{fi} ($i = 1$ to 5) in Btu/lb and the irreversibility in Btu/min for the cooling tower pictured. Assume $T_o = 90°F$, $p_o = 1$ atm, $\omega^o = 0.0153$ lb(v)/lb(a) ($\tilde{\omega}^o = 0.0246$, $x_v^o = 0.024$), $c_{pa} = 0.24$ Btu/lb°R, $c_{pv} = 0.45$ Btu/lb°R. See Fig. E6-3.

Solution. Neglecting all stray heat transfers and kinetic and potential energy changes, an availability equation for the control volume at steady-state gives

$$\dot{I} = \dot{m}_1(a_{f1} - a_{f4}) + \dot{m}_a(a_{f2} - a_{f5}) + \dot{m}_3 a_{f3}$$

---
[3] Recall from Sec. 5-2 that an availability diagram constructed for one dead state is not likely to be accurate for any other.

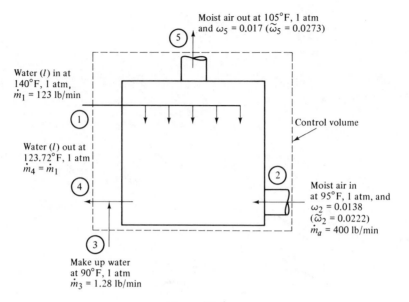

Figure E6-3

In this equation, a_{f2} and a_{f5} account, respectively, for the flow availability of the incoming and exiting moist air streams *per unit mass of dry air*. These are evaluated with Eq. (6-5b). The quantities a_{f1}, a_{f3}, and a_{f4} refer to the flow availability of the liquid only streams and are evaluated with Eq. (6-4e).

Inserting numerical values into Eq. (6-5b)

$$a_{f2} = (550)(0.24 + (0.0138)(0.45))\left[\frac{555}{550} - 1 - \ln\left(\frac{555}{550}\right)\right]$$

$$+ \left(\frac{1.986}{28.967}\right)(550)\left[(1.0222)\ln\left(\frac{1.0246}{1.0222}\right) + (0.0222)\ln\left(\frac{0.0222}{0.0246}\right)\right]$$

$$= 0.0056 + 0.0045 = 0.01 \text{ Btu/lb}$$

Similarly, $a_{f5} = 0.055$ Btu/lb.

To find a_{f1}, substitution into Eq. (6-4e) gives

$$a_{f1} = (107.96 - 1100.7) + (0.01629)(14.7 - 2.892)(144/778)$$

$$- 550\left[0.1985 - 2.0083 + \left(\frac{1.986}{18.106}\right)\ln\left(\frac{(0.024)(14.7)}{0.6988}\right)\right]$$

$$= 44.1 \text{ Btu/lb}$$

Similarly, $a_{f3} = 42$ Btu/lb and $a_{f4} = 42.5$ Btu/lb.

Finally, substituting into the expression for the irreversibility

$$\dot{I} = 123(44.1 - 42.5) + 400(0.01 - 0.055) + 1.28(42)$$
$$= 232.6 \text{ Btu/min}$$

6-6 Closure

This chapter extended the availability concept to include the work that in principle can be realized as the substances in a control mass under consideration are allowed to pass into, but not react chemically with, the environment. An important underlying assumption is that *each* of the substances contained within the control mass is *also* found within the environment. The next chapter considers the case where one or more of the control mass substances is not found in the environment. The availability concept is extended in such instances by including the work realizable through chemical reaction.

Selected References

1. EL-SAYAD, Y. M., and R. B. EVANS, "Thermoeconomics and the Design of Heat Systems," *ASME J. Engr. for Power*, **92**, January 1970, 27–35.
2. HAYWOOD, R. W., "A Critical Review of the Theorems of Thermodynamic Availability, with Concise Formulations," *J. Mech. Engr. Sci.*, **16**, *3*, 1974, 160–173 and *4*, 1974, 258–267.
3. WEPFER, W. J., R. A. GAGGIOLI, and E. F. OBERT, "Proper Evaluation of Available Energy for HVAC," *ASHRAE TRANS.*, **85** (Part 1), 1979, 214–229.

Problems

6-1 Develop Eq. (i) (page 128) of Sec. 6-2.
6-2 Figure P6-2 shows an ideal gas mixture containing substance k at temperature T and pressure p, separated from a gas phase of pure k at temperature T and pressure p' by a membrane permeable only to k. Assuming the ideal gas model for the pure

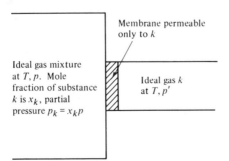

Figure P6-2

gas phase, show that for chemical equilibrium the pressure p' equals the partial pressure p_k $(= x_k p)$ of substance k in the mixture.

6-3 Develop Eq. (e) (page 131) of Sec. 6-3.

6-4 Show that Eq. (6-3c) is the minimum work *input* required to extract the mixture at the restricted dead state from the substances within the environment.

Base the development on the extraction unit shown in Fig. P6-4 which consists of n subunits, one for each substance, of which k is typical. For minimum work input each subunit necessarily operates reversibly. Central to its operation is a membrane permeable only to substance k.

Begin the development by writing an energy equation giving the work *input* to subunit k per mole of substance k

$$\bar{W}_k = -\bar{Q}_k - \bar{H}_k^o + \bar{H}_{k_o}$$

where $\bar{H}_k^o$ is the partial molal enthalpy of k within the environment and $\bar{H}_{k_o}$ is the partial molal enthalpy of k within the mixture at the restricted dead state. Next, write an entropy equation. Combine the two equations, use Eq. (6-2h), and reduce.

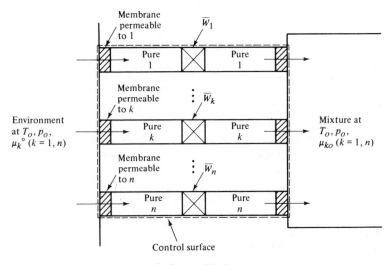

Figure P6-4

6-5 Develop Eq. (g) (page 133) of Sec. 6-3 by combining the steady-state forms of the energy and entropy equations, and using $d\bar{h} = T d\bar{s} + \bar{v} dp$ to reduce the result.

6-6 Develop Eq. (6-5b).

6-7 Referring to Problem 2-9 let gas 1 be 1 lbmol of O_2 at $60°F$ and 1 atm and gas 2 be 1 lbmol of N_2 at $60°F$ and 1 atm. For an environment consisting of dry air ($x_{O_2} = 0.21$, $x_{N_2} = 0.79$) at $60°F$ and 1 atm
(a) Determine the availability of the oxygen and of the nitrogen when separate.
(b) Determine the availability of the mixture after complete mixing has taken place.
(c) Using the results of parts (a) and (b) determine the irreversibility for the adiabatic mixing process.
(d) Determine the irreversibility using the results of Problem 2-9 and $I = T_o \sigma$.

Problems

6-8 Consider the adiabatic mixing of two moist air streams as shown in Fig. P6-8. Determine the flow availability in Btu/min at ports 1, 2, and 3, and the irreversibility in Btu/min. Take as the environment a mixture of dry air and water vapor at 95°F, 1 atm, $\omega° = 0.01406$. Let $c_{pa} = 0.24$ Btu/lb°R, $c_{pv} = 0.44$ Btu/lb°R. This case, as well as that of Problem 6-9, are among several discussed in Ref. [3].

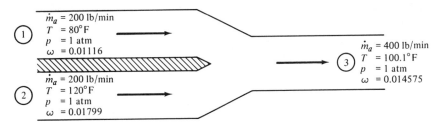

Figure P6-8

6-9 Repeat Problem 6-8 for the steam spray humidification device shown in Fig. P6-9.

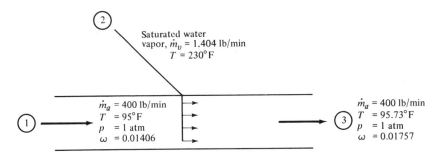

Figure P6-9

Chapter 7

Fuel Chemical Availability

The development of the chemical availability concept started in Chap. 6 is continued in this chapter. Considered here is the contribution to the availablity magnitude associated with the work that can be done by a combined system of control mass and environment as one or more of the substances making up the control mass at the restricted dead state react chemically with substances drawn from the environment to produce other environmental substances.

7-1 Introduction

To distinguish the presentation of the current chapter from those which have gone on before, consider the special case of a unit mass of a hydrocarbon fuel, C_aH_b, at the restricted dead state and let the *environment* consist of a gas phase involving oxygen, carbon dioxide, water vapor, and possibly other substances, but *not* the fuel. Since the fuel is at T_o, p_o, no thermomechanical availability is associated with it. Moreover, since the fuel is not a component of the environment, there is no chemical availability contribution of the type considered in Chap. 6. On the other hand, it is easy to imagine that by allowing the fuel to react with oxygen drawn from the environment considerable work can be done. This is the aspect of the chemical availability concept that is the focus of the current chapter.

In this chapter, the chemical availability concept is extended to include consideration of the work that can be developed as one or more of the substances making up a control mass at the restricted dead state react with substances drawn from the environment to produce other environmental substances. This case is important in practice because it applies to fuels.

7-2 Combustion Background

The developments of this chapter make use of some elementary ideas about chemical reactions, and in particular *combustion*. Since this topic is not normally treated in a first course in thermodynamics, background material related to it is presented in the next section before continuing the development of the availability concept in Sec. 7-3.

7-2 Combustion Background

In a chemical reaction the interatomic bonds in the molecules of the *reactants* are broken and the atoms and electrons rearranged to form *products*. One particular kind of reaction is emphasized here because of its importance in power production. This is *combustion*. Combustion is the rapid reaction between a fuel and oxygen in which chemical internal energy is liberated (exothermic reaction).

Associated with every chemical reaction is a chemical equation. For example, a reaction between methane and air to form carbon dioxide and water is expressed by the equation

$$CH_4 + 2(O_2 + 3.76N_2) \longrightarrow CO_2 + 2H_2O + 7.52N_2 \tag{a}$$

In writing Eq. (a), all components of air other than oxygen have been lumped together with nitrogen, and the molar ratio of nitrogen to oxygen taken to be 3.76. Also, the nitrogen is assumed to be inert, it does not combine with any other substance present. The numerical coefficients in the equation which precede the chemical symbols to give an equal number of atoms of each chemical element on both sides of the equation are called *stoichiometric coefficients*.

A combustion reaction of a hydrocarbon fuel is described as *complete* if all the carbon present is converted into CO_2 and all the hydrogen is converted into H_2O. The amount of air which supplies just enough oxygen for complete combustion is called the *theoretical amount of air*. Normally, the amount of air actually supplied is either less than or greater than the theoretical amount. Equation (a) describes the complete combustion of methane with the theoretical amount of air.

Enthalpy of Formation. Consider a reaction equation describing the combustion of hydrocarbon fuel C_aH_b with air

$$C_aH_b + \alpha(O_2 + 3.76N_2) \longrightarrow \sum_k v_k \pi_k \tag{b}$$

In this equation, π_k symbolizes products and the v_k are their stoichiometric coefficients, while α is the number of moles of oxygen participating in the reaction. The equation is written on a per mole of fuel basis.

Neglecting the effect of kinetic and potential energies, an energy equation *per mole of fuel consumed* for a control volume at steady-state is

$$0 = \bar{Q} - \bar{W} + \bar{H}_R - \bar{H}_P \qquad \text{(c)}$$

where

$$\begin{aligned} \bar{H}_R &= (\bar{h})_{\text{fuel}} + \alpha(\bar{h})_{O_2} + 3.76\alpha(\bar{h})_{N_2} \\ \bar{H}_P &= \sum_k \nu_k (\bar{h})_k \end{aligned} \qquad \text{(d)}$$

In Eqs. (d), $(\bar{h})_k$ is the enthalpy of substance π_k per mole, $(\bar{h})_{\text{fuel}}$ is the enthalpy of the fuel per mole of fuel, and so on.

An analysis using Eqs. (c) and (d) requires values for the enthalpies. When these are sought, it is found that for many pure substances they are not available on a common basis. That is, a datum state is used for each substance independent of that selected for any other (Sec. 2-4). These enthalpy values may be used in any analysis involving a single substance or a mixture of nonreacting substances because differences in property values are all that enter into consideration, and the effect of the arbitrary datum cancels. But when there is a chemical reaction, reactants disappear and products are formed, and differences can no longer be calculated for all substances involved. For this reason it is necessary to use a common basis when evaluating the enthalpies required in combustion applications.

A consistent and convenient basis can be established by assigning arbitrarily a value of zero to the enthalpy of the *elements* at a reference state of one atmosphere and 298.15 K (25°C). The enthalpy of any *compound* at 298.15 K and 1 atm is then equal to its *enthalpy of formation*, $\bar{h}_F$, at that state. The enthalpy of formation is the energy released or absorbed when the compound is formed from its elements, the compound and elements all being at 298.15 K, 1 atm. The enthalpy of formation is deduced from laboratory measurements of energy transfers or by statistical mechanical methods. Appendix Table B-7 gives the values of the enthalpy of formation $\bar{h}_F$ for a number of substances.

The enthalpy of a compound at any state other than the reference state is found by adding the change in enthalpy between the reference state and the given state to the enthalpy of formation

$$\bar{h}(T, p) = \bar{h}_F + \Delta\bar{h}$$

where, for $T_{\text{ref}} = 298.15$ K, $p_{\text{ref}} = 1$ atm

$$\Delta\bar{h} = [\bar{h}(T, p) - \bar{h}(T_{\text{ref}}, p_{\text{ref}})]$$

In evaluating $\Delta\bar{h}$, the choice of a datum is arbitrary because a difference is being determined. Accordingly, $\Delta\bar{h}$ may be found by use of tabular data such as in Appendix Tables B-2, B-3, B-4, and B-6, the generalized enthalpy departure chart Appendix Fig. B-2, or other means.

7-2 Combustion Background

Example 7-1. Consider the complete combustion of methane with 140% of the theoretical amount of air. Assuming the combustion products form an ideal gas mixture, determine their temperature, T_p, at the reactor exit. Let the reactor be at steady-state and well-insulated. There is no shaft work and kinetic and potential energy effects are negligible. See Fig. E7-1.

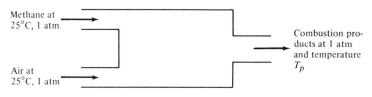

Figure E7-1

Solution. Equation (a) describes the reaction of methane with the theoretical amount of air. For 140% of the theoretical amount of air, the appropriate equation for complete combustion is

$$CH_4 + 2.8(O_2 + 3.76N_2) \longrightarrow CO_2 + 2H_2O(v) + 10.53N_2 + 0.8O_2$$

An energy equation reduces to $\bar{H}_P = \bar{H}_R$, where

$$\bar{H}_P = \{\bar{h}_F + [\bar{h}(T_p) - \bar{h}(T_{ref})]\}_{CO_2} + 2\{\bar{h}_F + [\bar{h}(T_p) - \bar{h}(T_{ref})]\}_{H_2O(v)}$$
$$+ 10.53\{\bar{h}_F + [\bar{h}(T_p) - \bar{h}(T_{ref})]\}_{N_2} + 0.8\{\bar{h}_F + [\bar{h}(T_p) - \bar{h}(T_{ref})]\}_{O_2}$$

$$\bar{H}_R = (\bar{h}_F)_{fuel} + 2.8\{\bar{h}_F + [\bar{h}(T_A) - \bar{h}(T_{ref})]\}_{O_2} + 10.53\{\bar{h}_F + [\bar{h}(T_A) - \bar{h}(T_{ref})]\}_{N_2}$$

In the expression for $\bar{H}_R$, T_A is the temperature of the air. For the case under consideration, $\bar{H}_R$ reduces to the enthalpy of formation of the fuel because $T_A = T_{ref}$, and $(\bar{h}_F)_{O_2} = (\bar{h}_F)_{N_2} = 0$ (Appendix Table B-7).

The remaining enthalpy of formation terms are found from Appendix Table B-7,

$$(\bar{h}_F)_{fuel} = -32{,}210 \text{ Btu/lbmol}$$
$$(\bar{h}_F)_{CO_2} = -169{,}300 \text{ Btu/lbmol}$$
$$(\bar{h}_F)_{H_2O(v)} = -104{,}040 \text{ Btu/lbmol}$$

Collecting these results, the energy equation reduces to

$$[\bar{h}(T_p) - \bar{h}(T_{ref})]_{CO_2} + 2[\bar{h}(T_p) - \bar{h}(T_{ref})]_{H_2O(v)} + 10.53[\bar{h}(T_p) - \bar{h}(T_{ref})]_{N_2}$$
$$+ 0.8[\bar{h}(T_p) - \bar{h}(T_{ref})]_{O_2} = 345{,}170 \text{ Btu/lbmol}$$

In this equation, the terms $\bar{h}(T_{ref})$ are found directly from Appendix Table B-6 at $T_{ref} = 298.15$ K (25°C). The only unknown is T_p. It can be determined in an iterative procedure as follows. The value of T_p is guessed. This enables the $\bar{h}(T_p)$ value for each of the products to be determined from Appendix Table B-6; with these, the left side of the equation is evaluated. The procedure continues until the value determined approximates the right side to the desired accuracy. The result of this procedure is 3371°R.

The *enthalpy of combustion* is the difference between the enthalpy of the products and the enthalpy of the reactants, $\bar{H}_P - \bar{H}_R$, when complete combustion occurs and both reactants and products are at the same temperature and pressure. For hydrocarbon fuels the enthalpy of combustion is negative in value since chemical internal energy is liberated in the reaction.

The *heating value* of a fuel is a positive number equal to the magnitude of the enthalpy of combustion. Two heating values are recognized by name: the *higher* heating value (HHV) and the *lower* heating value (LHV). The higher heating value is the value determined when the water formed during combustion is entirely a liquid. The lower heating value is the value when the water formed is entirely a vapor. Heating values for a number of hydrocarbons are listed in Table 7-2.

Absolute Entropy. An entropy equation, per mole of fuel, written for the reaction Eq. (b) and a control volume at steady-state is

$$0 = \sum_j \frac{\bar{Q}_j}{T_j} + \bar{S}_R - \bar{S}_P + \bar{\sigma}$$

where

$$\bar{S}_R = (\bar{s})_{\text{fuel}} + \alpha(\bar{s})_{O_2} + 3.76\alpha(\bar{s})_{N_2}$$
$$\bar{S}_P = \sum_k v_k(\bar{s}_k)$$

and $\bar{\sigma}$ accounts for the entropy production.

As for the case of enthalpy just considered, when entropy values are inserted in these equations care must be taken that the values are relative to a common basis. This is accomplished by means of the *third law of thermodynamics* as follows.

The empirically based *third law* states that the entropy of a pure crystalline substance is zero at the absolute zero of temperature, 0 K or 0°R. This provides a datum relative to which the entropy of each substance can be evaluated. The entropy relative to this datum is called the *absolute* entropy. The change in entropy of a substance between absolute zero and any given temperature can be determined either from experimental measurement of energy transfers or by procedures based on statistical mechanics.

Appendix Table B-7 gives the absolute entropy at 298.15 K and 1 atm for several substances. Appendix Table B-6 gives tabulations of absolute entropy for a number of gases versus temperature at a pressure of 1 atm. For the gases, ideal gas behavior is assumed in these tables.

With the value of the absolute entropy known at $T_{ref} = 298.15$ K, $p_{ref} = 1$ atm, the absolute entropy at any other state is found from

$$\bar{s}(T, p) = \bar{s}(T_{ref}, p_{ref}) + \Delta\bar{s} \tag{e}$$

where

$$\Delta\bar{s} = [\bar{s}(T, p) - \bar{s}(T_{ref}, p_{ref})]$$

In evaluating $\Delta\bar{s}$ the choice of a datum is arbitrary because a difference is being determined. Accordingly, it may be found by use of tabular entropy data such as in Appendix Tables B-2, B-3, B-4, and B-6, the generalized entropy departure chart Fig. B-3, or other means.

Gibbs Function. When the Gibbs function is used in combustion applications, there must also be due regard to the question of a datum. The procedure followed in setting a datum is like that used in defining the enthalpy of formation: To each element at the standard reference state of 298.15 K (25°C) and 1 atm is arbitrarily assigned a zero value of the Gibbs function. Then, when a compound is formed from its elements, the *Gibbs function of formation*, $\bar{g}_F$, of the compound is equal to the change in the Gibbs function for the reaction. Appendix Table B-7 gives the Gibbs function of formation of several substances.

The Gibbs function of a compound at any state other than the reference state is found by adding the change in Gibbs function between the reference state and the given state to the Gibbs function of formation

$$\bar{g}(T, p) = \bar{g}_F + \Delta\bar{g} \tag{7-2a}$$

where, for $T_{ref} = 298.15$ K, $p_{ref} = 1$ atm

$$\Delta\bar{g} = [\bar{g}(T, p) - \bar{g}(T_{ref}, p_{ref})]$$

With $\bar{g} = \bar{h} - T\bar{s}$, the last equation can be written as

$$\Delta\bar{g} = [\bar{h}(T, p) - \bar{h}(T_{ref}, p_{ref})] - [T\bar{s}(T, p) - T_{ref}\bar{s}(T_{ref}, p_{ref})] \tag{7-2b}$$

7-3 Fundamentals

This section extends the chemical availability concept by considering the work that can be developed as one or more of the substances making up a control mass at the restricted dead state react with substances drawn from the environment to produce other environmental substances. The methodology developed is generally applicable. However, to ease the introduction to this subject, initial emphasis is on the easily visualized case of fuels.

Derivation. Figure 7-1 shows a substance at the restricted dead state consisting of a fuel or a mixture in which it is a component. Also shown is a device which in principle permits work to be extracted as the fuel reacts with components from the environment to produce other environmental components. For maximum work the device necessarily operates reversibly. Semipermeable membranes, each of which permit only a particular substance to pass through, are important for its operation. The operation of this device is similar to that of such idealized devices as a reversible *fuel cell* or a *van't Hoff equilibrium box*.

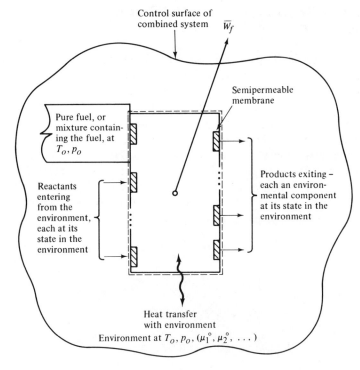

Figure 7-1 Reaction of fuel with environmental components.

Let the equation describing the reaction be

$$C_f + \sum_R v_i C_i \longrightarrow \sum_P v_\lambda C_\lambda \qquad (a)$$

where C_f is the chemical symbol of the fuel, C_i are the environmental reactants and v_i their stoichiometric coefficients per mole of fuel, C_λ are the environmental products and v_λ their stoichiometric coefficients.

For steady operation, an energy equation gives the work developed per mole of fuel, $\bar{W}_f$

$$\bar{W}_f = \bar{Q}_f + \bar{H}_{fo} + \sum_R v_i \bar{H}_i^\circ - \sum_P v_\lambda \bar{H}_\lambda^\circ$$

7-3 Fundamentals

where $\bar{H}_{fo}$ is the partial molal enthalpy of the fuel at the restricted dead state, $\bar{H}_i^o$ is the partial molal enthalpy of environmental reactant C_i, and so on.

An entropy equation gives after rearrangement

$$\bar{Q}_f = T_o(\sum_P v_\lambda \bar{S}_\lambda^o - \sum_R v_I \bar{S}_I^o - \bar{S}_{fo})$$

where $\bar{S}$ denotes a partial molal entropy.

Eliminating $\bar{Q}_f$ between the last two equations

$$\bar{W}_f = (\bar{H}_{fo} - T_o \bar{S}_{fo}) + \sum_R v_I(\bar{H}_I^o - T_o \bar{S}_I^o) - \sum_P v_\lambda(\bar{H}_\lambda^o - T_o \bar{S}_\lambda^o)$$

or, with Eq. (6-2h), in terms of chemical potentials

$$\bar{W}_f = \mu_{fo} - (\sum_P v_\lambda \mu_\lambda^o - \sum_R v_I \mu_I^o) \tag{b}$$

This is the contribution of the fuel to the chemical availability associated with the mixture at the restricted dead state.

If the fuel exists alone at the restricted dead state, Eq. (b) gives the *fuel chemical availability*

$$\bar{a}_f^{ch}(T_o, p_o) = \bar{g}_f(T_o, p_o) - (\sum_P v_\lambda \mu_\lambda^o - \sum_R v_I \mu_I^o) \tag{7-3a}$$

since for a pure substance, $\mu = \bar{g}$ (Sec. 6-2). The use of this result is amply illustrated in the current section.

It is frequently convenient to identify the bracketed term of Eq. (7-3a), and Eq. (b), as μ_f^o. In this way a chemical potential can be associated with the fuel *as if* it exists in the environment.[1]

$$\mu_f^o = \sum_P v_\lambda \mu_\lambda^o - \sum_R v_I \mu_I^o \tag{7-3b}$$

Introducing Eq. (7-3b) into Eq. (b) permits the contribution of the fuel to the mixture chemical availability to be written in the same form as the case treated in Chap. 6, as a difference in chemical potentials

$$\bar{W}_f = \mu_{fo} - \mu_f^o \tag{7-3c}$$

Using Eq. (7-3b) in Eq. (7-3a) and rearranging

$$\mu_f^o = \bar{g}_f(T_o, p_o) - \bar{a}_f^{ch}(T_o, p_o) \tag{7-3d}$$

Equation (7-3d) is useful for certain formulations of the availability equation, as illustrated in Sec. 7-5.

---

[1] Or, following Ref. [1], the chemical potential in the environment, μ_f^o, of a substance C_f, which is *not* among the substances making up the environment, can be derived from the equilibrium condition applying to the reaction of formation of C_f from environmental substances.

Hydrocarbon Fuels. The objective of this section is to show how the chemical availability of hydrocarbon fuels can be evaluated. For purposes of illustration, the environment is modeled as shown in Table 7-1. It is assumed to consist of a gas phase including water vapor and a number of dry components. The gas phase forms an ideal gas mixture.

Table 7-1

Environment Used to Evaluate the Chemical Availability of Hydrocarbon Fuels Reported in Table 7-2

$T_o = 298.15$ K (77°F), $p_o = 1$ atm

Gas Phase:	Substance	Mole Fraction (x^o)
	N_2	0.7567
	O_2	0.2035
	H_2O	0.0303
	CO_2	0.0003
	Other	0.0092

Consider a hydrocarbon fuel C_aH_b which is reacted with oxygen from the environment to produce carbon dioxide and water vapor as follows

$$C_aH_b + (a + b/4)O_2 \longrightarrow aCO_2 + b/2\, H_2O(v)$$

For C_aH_b this is the counterpart to Eq. (a).

Applying Eq. (7-3a)

$$\bar{a}_f^{ch} = \bar{g}_f(T_0, p_0) + (a + b/4)\mu_{O_2}^o - a\mu_{CO_2}^o - (b/2)\mu_{H_2O(v)}^o \qquad (c)$$

In the environment, the oxygen, carbon dioxide, and water vapor are components of an ideal gas mixture; thus, with Eq. (6-2f)

$$\mu_{O_2}^o = \bar{g}_{O_2}(T_o, p_o) + \bar{R}T_o \ln(x_{O_2}^o)$$

$$\mu_{CO_2}^o = \bar{g}_{CO_2}(T_o, p_o) + \bar{R}T_o \ln(x_{CO_2}^o)$$

$$\mu_{H_2O(v)}^o = \bar{g}_{H_2O(v)}(T_o, p_o) + \bar{R}T_o \ln(x_{H_2O}^o)$$

In these equations $x_{O_2}^o$, $x_{CO_2}^o$, $x_{H_2O}^o$ denote, respectively, the mole fraction of oxygen, carbon dioxide, and water vapor within the environment.

Collecting the last four expressions

$$\bar{a}_f^{ch}(T_o, p_o) = -\Delta G(T_o, p_o) + \bar{R}T_o \ln\left[\frac{(x_{O_2}^o)^{a+b/4}}{(x_{CO_2}^o)^a (x_{H_2O}^o)^{b/2}}\right] \qquad (7\text{-}3e)$$

where

$$\Delta G(T_o, p_o) = [a\bar{g}_{CO_2} + (b/2)\bar{g}_{H_2O(v)} - \bar{g}_f - (a + b/4)\bar{g}_{O_2}](T_o, p_o) \qquad (7\text{-}3f)$$

The chemical availability calculated with Eqs. (7-3e and f), relative to the

7-3 Fundamentals

environment of Table 7-1, for each of several pure hydrocarbons is reported in Table 7-2. The procedure is illustrated in Example 7-2.

Table 7-2

Chemical Availability of Selected Hydrocarbons C_aH_b at T_o, p_o

Values in kcal/gmol relative to the environment of Table 7-1[b]
$T_o = 298.15$ K, $p_o = 1$ atm

Fuel	Lower Heating Value (LHV)	Higher Heating Value (HHV)	$-\Delta G$[a]	$\bar{a}^{ch}$
Hydrogen H_2	57.8	68.32	56.69	56.22
Carbon(s) C	94.05	94.05	94.26	98.12
Paraffin Family C_nH_{2n+2}				
Methane (g) CH_4	191.76	212.80	195.50	198.42
Ethane (g) C_2H_6	341.27	372.82	350.73	357.04
Propane (g) C_3H_8	488.53	530.60	503.93	513.62
Butane (g) C_4H_{10}	635.39	687.98	656.74	669.82
Pentane (g) C_5H_{12}	782.06	845.16	809.48	825.96
Pentane (l) C_5H_{12}	775.70	838.80	809.23	825.71
Hexane (g) C_6H_{14}	928.95	1002.57	962.44	982.31
Hexane (l) C_6H_{14}	921.39	995.01	961.48	981.35
Heptane (g) C_7H_{16}	1075.87	1160.01	1115.43	1138.69
Heptane (l) C_7H_{16}	1067.13	1151.27	1113.76	1137.02
Octane (g) C_8H_{18}	1222.79	1317.45	1268.43	1295.07
Octane (l) C_8H_{18}	1212.87	1307.53	1266.06	1292.70
Olefin Family C_nH_{2n}				
Ethylene (g) C_2H_4	316.20	337.23	318.18	324.96
Propylene (g) C_3H_6	460.44	491.99	467.81	477.98
Butene (g) C_4H_8	607.69	649.76	621.02	634.58
Pentene (g) C_5H_{10}	754.26	806.85	773.54	790.48
Napthene Family C_nH_{2n}				
Cyclopentane (g) C_5H_{10}	740.80	793.39	763.98	780.93
Cyclopentane (l) C_5H_{10}	729.83	782.42	762.34	779.29
Cyclohexane (g) C_6H_{12}	881.68	944.78	913.29	933.63
Cyclohexane (l) C_6H_{12}	873.77	936.87	912.09	932.43
Aromatic Family C_nH_{2n-6}				
Benzene (g) C_6H_6	757.54	789.09	766.62	788.37
Toluene (g) C_7H_8	901.52	943.59	915.81	940.95
Toluene (l) C_7H_8	901.29	943.36	916.34	941.48
Ethylbenzene (g) C_8H_{10}	1048.54	1101.13	1068.74	1097.27

[a]Value for *liquid* H_2O: $-\Delta G = [\bar{g}_f + (a + b/4)\bar{g}_{O_2} - a\bar{g}_{CO_2} - (b/2)\bar{g}_{H_2O(l)}]$. The value for *vapor* H_2O is obtained by subtracting $(b/2)[\bar{g}_{H_2O(v)} - \bar{g}_{H_2O(l)}]$, approximately $1.027b$ kcal/gmol.

[b]Property values from F.D. Rossini, D.D. Wagman, W.H. Evans, S. Levine, and I. Jaffe, *Selected Values of Chemical Thermodynamic Properties*, NBS Circular 500, United States Department of Commerce—National Bureau of Standards, Washington DC, February 1, 1952.

Example 7-2. Determine the chemical availability of pure carbon relative to the reference environment of Table 7-1.

Solution. The reaction of carbon with environmental oxygen to produce environmental carbon dioxide is

$$C + O_2 \longrightarrow CO_2$$

Applying Eqs. (7-3e and f)

$$\bar{a}_C^{ch} = -(\bar{g}_{CO_2} - \bar{g}_{O_2} - \bar{g}_C)(T_o, p_o) + \bar{R}T_o \ln\left(\frac{x_{O_2}^o}{x_{CO_2}^o}\right)$$

Inserting $\bar{g}_{CO_2} = -94.26$ kcal/gmol, $\bar{g}_{O_2} = \bar{g}_C = 0$,[2]

$$\bar{a}_C^{ch} = 94.26 + \left(\frac{1.986}{10^3}\right)(298.15) \ln\left(\frac{0.2035}{0.0003}\right)$$

$$= 98.12 \text{ kcal/gmol}$$

A special case of interest is when the fuel is a member of an ideal gas mixture at the restricted dead state with mole fraction x_f. To calculate the contribution of the fuel to the chemical availability of the mixture, use Eq. (6-2f) to write

$$\mu_{fo} = \bar{g}_f(T_o, p_o) + \bar{R}T_o \ln(x_f)$$

Introducing this into Eq. (b) and evaluating the other terms as above, the contribution of the fuel to the chemical availability of the mixture

$$= \bar{a}_f^{ch}(T_o, p_o) + \bar{R}T_o \ln(x_f) \qquad (7\text{-}3g)$$

where $\bar{a}_f^{ch}(T_o, p_o)$ is calculated via Eqs. (7-3e and f) (see Problem 7-5).

Approximations. Checking values in Table 7-2 leads to the conclusion that the chemical availability of a hydrocarbon fuel is within a few percent of the heating values and of $[-\Delta G(T_o, p_o)]$. These approximations make frequent appearances in the technical literature. Their use is illustrated in Examples 7-3 and 7-4.

Example 7-3. Consider a gas-fired home furnace which delivers a heat transfer $\bar{Q}$ (per mole of fuel consumed) at a temperature T_R. Let the heat transfer $\bar{Q}$ equal a fraction of the fuel's lower heating value: $\bar{Q} = \eta(\text{LHV})$, $\eta < 1$, and assume the remainder of the energy released upon combustion is lost up the stack and in miscellaneous leaks. (a) Develop the expression for the second law efficiency given below by Eq. (d). Approximate the availability of the fuel

[2] See note[b] in Table 7-2.

7-3 Fundamentals

consumed by its lower heating value, and neglect the availability input with the combustion air (if any). (b) Evaluate the efficiency for $\eta = 0.77$, $T_R = 580°R$, $T_o = 490°R$. Compare with the result of Example 4-5 and discuss.

Solution. (a) A second law efficiency can be formed as the ratio of the availability of the energy delivered at T_R to the availability of the fuel consumed (product/input)

$$\epsilon = \frac{\left(1 - \frac{T_o}{T_R}\right)\bar{Q}}{\bar{a}_f^{ch}}$$

With $\bar{Q} = \eta(\text{LHV})$ and $\bar{a}_f^{ch} \approx (\text{LHV})$, this reduces to

$$\epsilon = \eta\left(1 - \frac{T_o}{T_R}\right) \quad \text{(d)}$$

(b) Inserting values, $\epsilon = 0.12$ (12%).

Discussion. The low efficiency value calculated indicates that the fuel is not being utilized effectively. Relatively high potential fuel is used to produce only a relatively low potential heat transfer at temperature T_R. Better utilization is achieved in the heat pump system of Example 4-5 which delivers more energy for the same fuel expenditure. It should be noted, however, that heat pump systems are often more complex than furnace systems, and may well cost more to install and maintain.

The poor efficiency for the furnace is due mainly to the irreversible combustion process itself and not to losses, for even if $\eta = 100\%$, ϵ is improved only to about 16%. A more detailed look is taken at the irreversibility of combustion later in this section and in Sec. 7-4.

Example 7-4. Consider the steam generator shown in Fig. E7-4. Approximating the availability of the fuel consumed by its higher heating value, develop the second law efficiency expression given below by Eq. (e) and evaluate it for a steam generator heating efficiency $\eta = 70\%$. Discuss. Take $T_o = 520°R$.

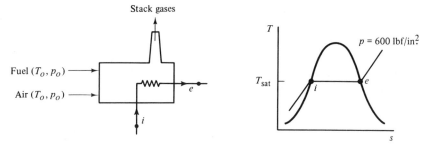

Figure E7-4

Solution. A *heating* efficiency for the steam generator can be written as

$$\eta = \frac{\dot{m}_s(h_e - h_i)}{\dot{E}_f}$$

where $\dot{E}_f$ equals the product of fuel flow rate and the fuel higher heating value, and $\dot{m}_s$ is the steam flow rate. A second law efficiency similar in form is

$$\epsilon = \frac{\dot{m}_s(a_{fe} - a_{fi})}{\dot{A}_f}$$

where $\dot{A}_f$ is the rate availability enters with fuel. By assumption, $\dot{A}_f \approx \dot{E}_f$. Combining these expressions

$$\epsilon = \eta \left(\frac{a_{fe} - a_{fi}}{h_e - h_i} \right)$$

$$= \eta \left[1 - \frac{T_o(s_e - s_i)}{(h_e - h_i)} \right]$$

For a saturated liquid at inlet i and saturated vapor at exit e, use of Eq. (2-4b) gives $(h_e - h_i) = T_{\text{sat}}(s_e - s_i)$, where T_{sat} is the saturation temperature. So, finally

$$\epsilon = \eta \left(1 - \frac{T_o}{T_{\text{sat}}} \right) \quad \text{(e)}$$

At $p = 600$ lbf/in.^{2}, $T_{\text{sat}} = 946°$R. With $\eta = 70\%$, $\epsilon = 32\%$.

Discussion. The efficiency η charges the device only for its energy *losses*. The efficiency ϵ charges it both for losses and the availability destruction which takes place due to internal irreversibilities. For two devices with the same value for η, Eq. (e) shows that the one with the higher saturation temperature (pressure) would have the higher second law efficiency. Attempts to improve ϵ for the steam generator might focus initially on improving its heating efficiency η. The theoretical limit is $\epsilon = 45\%$, corresponding to $\eta = 100\%$. Further improvement can come only by dealing directly with the irreversibilities inherent in the process itself.

Extension of the Method. As noted at the outset of this section, the method developed to determine the chemical availability of hydrocarbons is valid for other substances as well. This is illustrated in the next example for the case of carbon monoxide. Further discussion of this point is provided in Sec. 7-6.

Example 7-5. Consider carbon monoxide as a member of an ideal gas mixture at the restricted dead state with mole fraction x_{CO}. Determine its contri-

7-3 Fundamentals

bution to the chemical availability of the mixture, relative to the environment of Table 7-1.

Solution. Carbon monoxide can be reacted with environmental oxygen to produce environmental carbon dioxide.

$$CO + \tfrac{1}{2}O_2 \longrightarrow CO_2$$

Applying Eq. (b), the contribution of CO to the mixture's chemical availability

$$= (\mu_{CO,o} + \tfrac{1}{2}\mu^o_{O_2} - \mu^o_{CO_2})$$

where

$$\mu_{CO,o} = \bar{g}_{CO}(T_o, p_o) + \bar{R}T_o \ln(x_{CO})$$
$$\mu^o_{O_2} = \bar{g}_{O_2}(T_o, p_o) + \bar{R}T_o \ln(x^o_{O_2})$$
$$\mu^o_{CO_2} = \bar{g}_{CO_2}(T_o, p_o) + \bar{R}T_o \ln(x^o_{CO_2})$$

Collecting the last four equations, the contribution of the CO

$$= \left\{ (\bar{g}_{CO} + \tfrac{1}{2}\bar{g}_{O_2} - \bar{g}_{CO_2})(T_o, p_o) + \bar{R}T_o \ln\left[\frac{(x^o_{O_2})^{1/2}}{x^o_{CO_2}}\right] + \bar{R}T_o \ln x_{CO} \right\}$$

With,[3] $\bar{g}_{CO_2} = -94.26$ kcal/gmol, $\bar{g}_{CO} = -32.81$, $\bar{g}_{O_2} = 0$, $x^o_{O_2} = 0.2035$, $x^o_{CO_2} = 0.0003$, the contribution of CO to the mixture's chemical availability

$$= (65.78 + \bar{R}T_o \ln x_{CO}) \text{ kcal/gmol}$$

Combustion Irreversibility. To calculate the irreversibility, the only property of the environment that *must* be specified is the temperature T_o. This has been discussed in Sec. 3-7, illustrated in Example 6-1, and is also brought out by the following simple illustration involving the irreversibility associated with combustion.

Consider a combustor at steady-state in which carbon and oxygen, each pure at T, p enter and products of combustion at T', p' exit. The products form an ideal gas mixture. Let the reaction be

$$C + \tfrac{3}{4}O_2 \longrightarrow \tfrac{1}{2}CO_2 + \tfrac{1}{2}CO$$

For simplicity, assume the combustor operates adiabatically and with negligible kinetic and potential energy changes. The availability calculations are relative to an environment like that of Table 7-1 which includes oxygen and carbon dioxide, but not carbon or carbon monoxide.

[3] See note[b] in Table 7-2.

Using Eq. (6-4b) to evaluate the flow availabilities, the irreversibility per mole of carbon is

$$\bar{I} = [\bar{h}_C(T,p) - T_o\bar{s}_C(T,p) - \mu_C^o] + \tfrac{3}{4}[\bar{h}_{O_2}(T,p) - T_o\bar{s}_{O_2}(T,p) - \mu_{O_2}^o]$$
$$- \{\tfrac{1}{2}[\bar{h}_{CO_2}(T') - T_o\bar{s}_{CO_2}(T', x_{CO_2}p') - \mu_{CO_2}^o]$$
$$+ \tfrac{1}{2}[\bar{h}_{CO}(T') - T_o\bar{s}_{CO}(T', x_{CO}p') - \mu_{CO}^o]\} \tag{f}$$

In evaluating Eq. (f) a distinction must be made between substances that exist within the environment (O_2 and CO_2) and those which do not (C and CO). For the substances that do not exist within the environment, the chemical potentials required by Eq. (f) are determined with Eq. (7-3b)

$$\mu_C^o = \mu_{CO_2}^o - \mu_{O_2}^o$$
$$\mu_{CO}^o = \mu_{CO_2}^o - \tfrac{1}{2}\mu_{O_2}^o \tag{g}$$

Inserting Eqs. (g) into Eq. (f) and collecting like terms

$$\bar{I} = T_o[\tfrac{1}{2}\bar{s}_{CO_2}(T', x_{CO_2}p') + \tfrac{1}{2}\bar{s}_{CO}(T', x_{CO}p') - \bar{s}_C(T,p) - \tfrac{3}{4}\bar{s}_{O_2}(T,p)]$$
$$+ [\bar{h}_C(T,p) + \tfrac{3}{4}\bar{h}_{O_2}(T,p) - \tfrac{1}{2}\bar{h}_{CO_2}(T') - \tfrac{1}{2}\bar{h}_{CO}(T')]$$
$$+ [-(\mu_{CO_2}^o - \mu_{O_2}^o) - \tfrac{3}{4}\mu_{O_2}^o + \tfrac{1}{2}\mu_{CO_2}^o + \tfrac{1}{2}(\mu_{CO_2}^o - \tfrac{1}{2}\mu_{O_2}^o)]$$

The second term on the right side vanishes by application of an energy equation for the adiabatic combustor at steady-state. The third term vanishes identically. Accordingly, the expression reduces to $\bar{I} = T_o\bar{\sigma}$, where $\bar{\sigma}$ is the entropy production per mole of carbon.

This illustrates that to perform calculations aimed at determining the irreversibility it is not necessary to specify either the pressure p_o or the chemical potentials. The only environmental property that *must* be specified is the temperature T_o.

7-4 Application to a Vapor Power Plant

Introduction. The purpose of this section is to illustrate various aspects of availability analysis through the study of an idealized power plant in which fuel is burned, vapor is generated and power developed. Figure 7-2 shows a schematic of the power plant. Air and fuel enter and power is produced. Also exiting are stack gases, refuse, and miscellaneous heat transfers. As shown on the figure, cooling water enters at one point and leaves at another. The availability calculations are relative to the environment of Table 7-1.

An energy-based efficiency for the plant written in the form (product/input) is

7-4 Application to a Vapor Power Plant

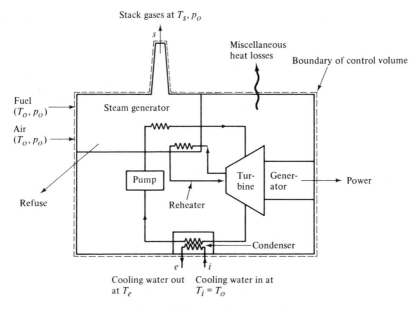

Figure 7-2 Power plant schematic.

$$\eta = \frac{\text{Power developed}}{\dot{E}_f}$$

$$= \frac{\text{Power developed}}{(\text{Fuel flow rate})(\text{Higher heating value})} \quad (a)$$

which is the ratio of the power developed to the limiting amount of chemical internal energy liberated in the combustion process. Values of the plant efficiency are commonly in the range 30 to 45%.

Regarding as losses the availability carried out with cooling water, stack gases, refuse, and the miscellaneous heat transfers, a second law efficiency in the form (product/input) is

$$\epsilon = \frac{\text{Power developed}}{\dot{A}_f}$$

$$= \frac{\text{Power developed}}{(\text{Fuel flow rate})(\bar{a}_f^{ch})} \quad (b)$$

Since $\bar{a}_f^{ch} \approx$ (HHV) (Table 7-2), the two efficiencies are seen to be roughly equal in value.

In the presentation to follow, the primary sources of availability losses and destructions associated with the power plant are identified and evaluated. The destructions are found to be greater in magnitude than the losses. It is also shown that the principal irreversibility is associated with the steam generator.

Equation (b) is another example of a *task* efficiency. Here the task is to produce power from fuel. The denominator represents the maximum theoretical amount that can be produced from a certain quantity of fuel by any means whatever, the numerator is the amount produced in the particular system under consideration. The 30 to 45% efficiency value of vapor power plants should not be regarded as defining a practical limit for the production of electricity from fossil fuel, for the fuel cell is a device that has a greater value for the conversion efficiency because it does not have the same irreversibilities associated with its operation.

Condenser. The objective of this subsection is to evaluate the *availability* carried out of the plant with the condenser cooling water. But first, for contrast, the *energy* carried out with the condenser water is considered.

If the plant efficiency is η, a fraction $(1 - \eta)$ of the energy released in combustion is carried out in some way other than developed power, for example with the cooling water, stack gases, miscellaneous heat transfers, etc. The dominant portion of this is due to the energy exiting with the cooling water: $\dot{m}_w(h_e - h_i)$. The cooling water flow rate is $\dot{m}_w$. Thus, the net rate *energy* is carried out with cooling water is roughly

$$\dot{m}_w(h_e - h_i) \approx (1 - \eta)\dot{E}_f$$

The net rate *availability* is carried out with cooling water is

$$\dot{A}_w = \dot{m}_w[(h_e - h_i) - T_o(s_e - s_i)]$$
$$= \dot{m}_w(h_e - h_i)\left[1 - T_o\frac{(s_e - s_i)}{(h_e - h_i)}\right]$$

Since $\dot{m}_w(h_e - h_i) \approx (1 - \eta)\dot{E}_f$

$$\frac{\dot{A}_w}{\dot{E}_f} \approx (1 - \eta)\left[1 - T_o\frac{(s_e - s_i)}{(h_e - h_i)}\right]$$

This expression is now rewritten as Eq. (c).

Regarding the condenser water as incompressible with an average specific heat c, and ignoring the condenser pressure drop

$$s_e - s_i = c \ln(T_e/T_i)$$
$$h_e - h_i = c(T_e - T_i)$$

Also, with the approximation $\bar{a}_f^{ch} \approx (\text{HHV})$, it follows that $\dot{E}_f \approx \dot{A}_f$.

Collecting results, the fraction of the availability entering the plant with the fuel that exits in the cooling water is approximately

$$\frac{\dot{A}_w}{\dot{A}_f} \approx (1 - \eta)\left[1 - \frac{T_o \ln(T_e/T_i)}{(T_e - T_i)}\right] \tag{c}$$

Taking for illustrative purposes, $T_i = T_o = 537°R$, $T_e = 565°R$, and $\eta = 40\%$

7-4 Application to a Vapor Power Plant

$$\frac{\dot{A}_w}{\dot{A}_f} \approx 0.015 (1.5\%)$$

That is, less than 2% of the availability entering the plant with the fuel is carried out in the condenser water. Thus, in spite of the fact that a significant amount of *energy* exits in the cooling water, failure to utilize the availability in the condenser water is seen to be not especially wasteful. Nonetheless, power plant engineers should continue to be alert for economical uses for this warm water.

Stack Gases. In this subsection the availability carried out with the stack gases is considered. It is assumed that methane is the fuel, it is burned with 140% of the theoretical amount of air, and combustion is complete. The reaction equation is[4]

$$CH_4 + 2.8(O_2 + 3.76\, N_2) \longrightarrow CO_2 + 2H_2O(v) + 10.53\, N_2 + 0.8\, O_2$$

The required availability calculations are detailed in Example 6-2.

It is shown in the solution to Example 6-2 that there are two contributions to the availability of the combustion products, one associated with composition and one associated with temperature. The compositional contribution is calculated to be 7636 Btu/lbmol. Taking for illustrative purposes $T_s = 740°R$ as the temperature at which the gases exit, the contribution associated with temperature is 3243 Btu/lbmol. For methane, $\bar{a}_f^{ch} = 357{,}130$ Btu/lbmol (from Table 7-2). Accordingly, the availability exiting in the combustion products amounts to 10,879 Btu/lbmol, which is about 3% of the availability entering with the fuel.

Steam Generator. Assuming for illustrative purposes that the second law efficiency ϵ for the plant is on the order of 40%, it follows that for every 100 units of availability entering with the fuel, 60 units are either carried out in various effluent streams or are destroyed due to irreversibilities within the plant components. According to the previous calculations, about 5 units are carried out with the cooling water and stack gases. This leaves a balance of 55 units that are destroyed or carried out in other ways. A small portion of this can be charged to effects such as miscellaneous heat transfer which have not been evaluated. The largest part of the balance, however, is due to irreversibilities within the plant components, and the steam generator is the principal contributor to this.

There are two main sources of irreversibility within the steam generator. One is the irreversible heat transfer that occurs between the hot combustion products and the fluid in the boiler tubes. This kind of irreversibility is considered in Example 4-3. The other is the irreversible combustion process itself. The significance of the combustion irreversibility can be estimated by idealizing

---

[4] To simplify, all components of the combustion air other then oxygen are lumped together with nitrogen. And to be consistent with the discussion of Sec. 7-2, the molar ratio of nitrogen to oxygen is taken as 3.76.

the steam generator as consisting of an adiabatic unit in which fuel and air are burned to produce hot combustion products, followed by a heat exchanger unit where the hot products are used to generate, superheat, and reheat steam.

An energy equation for the adiabatic combustion unit reduces to $\bar{H}_P = \bar{H}_R$. Using this for the reaction given above, the temperature of the combustion products is $T_p = 3371°R$ (see Example 7-1). Assuming the mixture pressure is 1 atm, the availability of the gases exiting the combustion unit is the sum of a contribution associated with composition and one associated with temperature. The compositional contribution is the same as for the stack gases: 7636 Btu/lbmol. The contribution related to temperature, evaluated at $T_p = 3371°R$, is 230,276 Btu/lbmol (see Example 6-2). The availability of the gases exiting the combustion unit is the sum, 237,912 Btu/lbmol, with the temperature contribution being by far the dominant one.

A second law efficiency for the combustion unit can be written as[5]

$$\epsilon = \frac{(\dot{A}_{\text{products}}/\dot{N}_{\text{fuel}}) - (\dot{A}_{\text{air}}/\dot{N}_{\text{fuel}})}{\bar{a}_f^{ch}} \tag{7-4a}$$

Inserting values, $\epsilon = 66\%$. In other words, for every 100 units of availability carried in with the fuel 34 units are destroyed in the combustion unit. The primary source of destruction is that high potential fuel is consumed in the spontaneous combustion process to produce relatively low temperature (low potential) combustion products.

Table 7-3 summarizes the calculations of this section.

Table 7-3

Summary of Availability Losses and Destructions

Availability in with fuel	100 units
Power developed	− 40
Balance	60
Availability out with cooling water	− 2
Availability out with stack gases	− 3
Balance	55
Destruction in combustion unit	− 34
Balance. (Destructions in various components: combustion heat exchanger, turbine, etc. plus miscellaneous losses. The destruction associated with heat transfer in the steam generator is the largest part.)	21

[5]This form is suggested by Z. Rant, "The Influence of Air Preheating on the Irreversibilities of Combustion," *Brennst. Warme-Kraft*, **13**, *11*, 1961, 496–500. (Translation available through the Central Technical Information Service of the Central Electricity Generating Board, London, England). If the combustion air is not *preheated*, it is assumed to enter with zero availability, $\dot{A}_{\text{air}}/\dot{N}_{\text{fuel}} = 0$.

7-4 Application to a Vapor Power Plant

Air Preheating. A well-known method to increase the efficiency of combustion is to preheat the combustion air. The improvement that can be realized in this way is shown in Fig. 7-3 which is a plot of the second law efficiency and combustion product temperature T_p versus the temperature T_A at which air enters the combustor. Values are shown for the adiabatic and complete combustion of methane with 100%, 140%, and 180% of the theoretical amount of air. The temperature T_p is determined as shown in Example 7-1. The second law efficiency is calculated by Eq. (7-4a), using the following equation, obtained by reducing Eq. (6-4f), to account for the fact that the air enters at temperature $T_A > T_o$

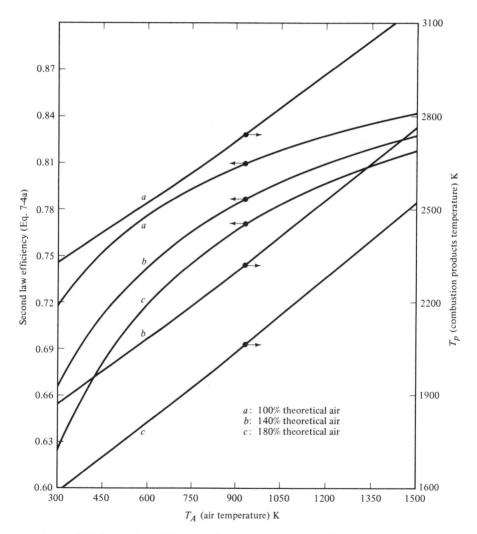

Figure 7-3 Second law efficiency and temperature of combustion products versus air temperature for complete combustion of methane.

$$\frac{\dot{A}_{air}}{\dot{N}_{fuel}} = 2.8[\bar{h}(T_A) - \bar{h}(T_o) - T_o(\bar{s}'(T_A) - \bar{s}'(T_o))]_{O_2}$$
$$+ 10.53[\bar{h}(T_A) - \bar{h}(T_o) - T_o(\bar{s}'(T_A) - \bar{s}'(T_o))]_{N_2}$$

Section Closure. The calculations of this section are intended to illustrate the application of availability principles through the analysis of an idealized vapor power plant. There are no fundamental difficulties, however, in applying the methodology to actual power plants where effects such as miscellaneous heat losses, infiltration of air, fuel mixtures (including water, sulfur, etc.), incomplete combustion, and so on must be accounted for.

7-5 Analysis of a Coal Gasification Reactor

Introduction. The purpose of this section is to illustrate the use of availability analysis through the study of an elementary coal gasification process.

There are several processes whereby coal can be converted to a combustible gas or liquid. The one under consideration here is the *carbon-steam* process. In the process coal is reacted with steam to produce hydrogen. The energy required for this highly *endothermic* process is provided by combustion of coal with air or oxygen.

Figure 7-4 shows the gasification reactor under consideration. It is assumed to operate at steady-state, with no heat losses, and with negligible kinetic and potential energy effects. The coal is idealized as pure carbon. The numerical coefficients before each chemical substance shown on the figure denote the moles of that substance *per mole of carbon* entering. The product gas is assumed to form an ideal gas mixture.

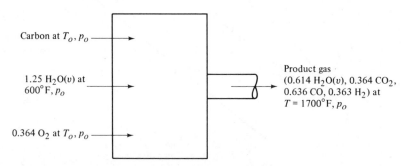

Figure 7-4 Gasification reactor.

Availability Analysis. The availability calculations to follow are relative to the environment of Table 7-1. First the reactants are considered and then the products. A summary of the results obtained is given in Table 7-4.

Pure carbon enters at temperature T_o and pressure p_o. The availability input is thus the chemical availability of carbon as given in Table 7-2: 98,120 kcal/

7-5 Analysis of a Coal Gasification Reactor

kgmol or 176,600 Btu/lbmol. Also, pure oxygen enters at T_o, p_o with availability (Eq. (6-3d))

$$= \bar{R}T_o \ln(1/x^o_{O_2})$$
$$= -(1.986)(537) \ln(0.2035)$$
$$= 1700 \text{ Btu/lbmol}$$

Steam enters at 600°F, 1 atm with availability (Eq. (6-4d))

$$= [\bar{h}(T,p) - \bar{h}_g(T_o)] - T_o[\bar{s}(T,p) - \bar{s}_g(T_o)] - \bar{R}T_o \ln(x^o_{H_2O}p_o/p_g(T_o))$$
$$= (18)[1335.2 - 1095.1 - 537(1.9737 - 2.044)]$$
$$- (1.986)(537) \ln((0.0303)(14.7)/0.46)$$
$$= 5040 \text{ Btu/lbmol}$$

Property values for the steam are obtained from Appendix Table B-2.

The product gas exiting at 1700°F, 1 atm is regarded as an ideal gas mixture. Making use of Eq. (6-4b), its flow availability *per mole of carbon* is determined from

$$(\dot{A}_{\text{product}}/\dot{N}_c) = \sum_{k=1}^{n} N_k[\bar{h}_k(T) - T_o \bar{s}_k(T, x_k p_o) - \mu^o_k]$$

where $\dot{N}_c$ is the molar flow rate of carbon and N_k is the number of moles of substance k per mole of carbon.

If substance k appears in the environment, μ^o_k can be eliminated from this expression by using Eq. (6-2e) to write

$$\mu^o_k = \bar{h}_k(T_o) - T_o \bar{s}_k(T_o, x^o_k p_o)$$

and so its contribution to the sum

$$= \{\bar{h}_k(T) - \bar{h}_k(T_o) - T_o[\bar{s}'_k(T) - \bar{s}'_k(T_o)] + \bar{R}T_o \ln(x_k/x^o_k)\}$$

This applies to the water vapor and the carbon dioxide.

If substance k does not appear in the environment (carbon monoxide, hydrogen), μ^o_k can be eliminated by using Eq. (7-3d) to write

$$\mu^o_k = \bar{g}_k(T_o, p_o) - \bar{a}^{ch}_k(T_o, p_o)$$

and so its contribution to the sum

$$= \{\bar{h}_k(T) - \bar{h}_k(T_o) - T_o[\bar{s}'_k(T) - \bar{s}'_k(T_o)] + \bar{R}T_o \ln(x_k) + \bar{a}^{ch}_k(T_o, p_o)\}$$

With these considerations, the flow availability of the product gas per mole of carbon is

$(\dot{A}_{\text{product}}/\dot{N}_c)$

$$= 0.614[\bar{h}(T) - \bar{h}(T_o) - T_o(\bar{s}'(T) - \bar{s}'(T_o)) + \bar{R}T_o \ln(x_{\text{H}_2\text{O}}/x^o_{\text{H}_2\text{O}})]_{\text{H}_2\text{O}(v)}$$
$$+ 0.364[\bar{h}(T) - \bar{h}(T_o) - T_o(\bar{s}'(T) - \bar{s}'(T_o)) + \bar{R}T_o \ln(x_{\text{CO}_2}/x^o_{\text{CO}_2})]_{\text{CO}_2}$$
$$+ 0.636[\bar{h}(T) - \bar{h}(T_o) - T_o(\bar{s}'(T) - \bar{s}'(T_o)) + \bar{R}T_o \ln(x_{\text{CO}})$$
$$+ \bar{a}^{ch}(T_o, p_o)]_{\text{CO}} + 0.636[\bar{h}(T) - \bar{h}(T_o) - T_o(\bar{s}'(T) - \bar{s}'(T_o))$$
$$+ \bar{R}T_o \ln(x_{\text{H}_2}) + \bar{a}^{ch}(T_o, p_o)]_{\text{H}_2}$$

As an illustration, the contribution of hydrogen to this sum

$$= \bar{h}(T) - \bar{h}(T_o) - T_o[\bar{s}'(T) - \bar{s}'(T_o)] + \bar{R}T_o \ln(x_{\text{H}_2}) + \bar{a}^{ch}(T_o, p_o)$$
$$= [15{,}158 - 3640] - 537[41 - 31.2] + (1.986)(537)\ln(0.2827) + 101{,}190$$
$$= 106{,}100 \text{ Btu/lbmol}$$

$\bar{a}^{ch}(T_o, p_o)$ comes from Table 7-2. The enthalpy and entropy values are from Appendix Table B-6G. Numerical results for the other products are listed in Table 7-4 (see Problem 7-7).

Table 7-4

Summary of the Availability Analysis for the Coal Gasification Reactor

	N_k	$T(°R)$	$p(atm)$	$\bar{a}_{fk}(Btu/lbmol)$	$N_k\bar{a}_{fk}(Btu)$
Reactants					
C	1	537	1	176,600	176,600
$H_2O(v)$	1.25	1060	1	5,040	6,300
O_2	0.364	537	1	1,700	620
				total:	183,520
Products					
$H_2O(v)$	0.614	2160	0.2729	10,560	6,480
CO_2	0.364	2160	0.1617	17,420	6,340
CO	0.636	2160	0.2827	123,790	78,730
H_2	0.636	2160	0.2827	106,100	67,480
				Total:	159,030

Referring to Table 7-4, it is seen that the contribution of the carbon accounts for the dominant portion of the availability entering the reactor. About 13% of this sum is destroyed due to irreversibilities within the reactor. Further, the hydrogen produced accounts for about 42% of the availability exiting. Considering that a significant fraction of the hydrogen availability will be destroyed in a subsequent combustion process, it is seen that the overall efficiency from carbon input to end use is not high.

7-6 Evaluation of Chemical Processes

Introduction. The environments used in previous developments to calculate availability magnitudes are adequate for the analysis of a wide range of practical applications, including ones involving combustion. But when other chemical processes are taken up, the question of how the environment is specified requires further consideration. In this section, some general ideas relative to this question are presented.

Though much has been written on how the environment is to be selected, there is agreement only on certain broad principles. Considerable discussion continues over details. Accordingly, before availability calculations involving chemical processes are carried out, it is wise to review the pertinent literature with the objective of selecting an approach which provides a suitable combination of accuracy and ease of use.

Modeling the Environment. There is no one theoretically *correct* environment. In many cases, the environment is determined through engineering intuition and arguments of utility. Considering the vast array of potential applications, it is not surprising that there is no one choice that suffices for all. There are, however, some general ideas which can guide the selection.

The environment is considered to be in stable equilibrium. Associated with it is a unique temperature, a unique pressure, and unique chemical potentials for the components making it up. These values do not change as a result of any of the processes under consideration. All substances of interest should be formable from the substances making up the environment.

It is advantageous to adopt as environmental components the most common ones as they occur naturally, since then their concentrations are known with a high degree of accuracy. By assigning zero availability to the most common ones, others have positive availabilities. These positive values are an index of their economic value (and, when in effluent streams, their influence on nature). Accordingly, the selection of environmental substances involves thermodynamic and economic judgments, at least implicitly.

The method used in Sec. 7-3 to determine the chemical availability of hydrocarbon fuels relative to an environment such as the one in Table 7-1 is valid for other substances as well. This is illustrated in Example 7-5 for the case of carbon monoxide, whose *stable* configuration within the Earth and its atmosphere is carbon dioxide, which is one of the environmental substances included in Table 7-1. To handle other substances of interest, the list of environmental components must be extended. For example, in Ref. [3] gypsum, $CaSO_4 \cdot 2H_2O$, is included as the stable form of sulfur, and calcite, $CaCO_3$, as the stable form of calcium (see Problem 7-15). This procedure can be followed for each substance of interest, by seeking its eventual form when released into the Earth and its atmosphere.

An important consideration in including a particular substance within the environment used for availability calculations is that it is the stable form of some substance of interest. Another consideration, which can be overriding, is the accessibility of the substance in the region where the plant or device is to operate. For example, if the production of methanol from natural gas in a desert area is under consideration, the inclusion of abundant liquid water in the environment may not be justified. Or, if only saline water or polluted water is available at some location, it may be desirable to regard pure water as having a positive availability.

Standard Chemical Availabilities. Because of the variation in conditions from place to place, the specification of a reference environment suitable for a given application can necessitate extensive deliberation and study. Furthermore, once the environment is decided upon, a series of calculations is required to obtain availability values for the substances of interest. These complexities can be sidestepped by use of "standard" chemical availabilities.

Standard chemical availabilities are determined relative to a "standard" environment. This permits the development of a table of standard chemical availabilities. Use of such a table greatly facilitates the application of availability analyses since it eliminates the need for many intermediate calculations. However, the term "standard" is somewhat misleading, for as noted previously there is no one specification of the environment that suffices for all applications.

In one approach, described in Ref. [6], standard chemical availabilities are calculated using the assumption that the environmental pressure and temperature have standard values and the environment consists of a number of reference substances, one for each chemical element, with standard concentrations based on the average concentration in the natural environment. The reference substances selected fall into three groups: gaseous components of the atmosphere, solid substances from the lithosphere, and both ionic and nonionic substances from the oceans. Table 7-5 gives standard chemical availabilities for certain of the elements as reported in Ref. [6]. The standard chemical availabilities for compounds are calculated relative to the standard chemical availabilities for the elements (see Problem 7-16). It is claimed that the departure of the actual environment from the standard environment only occasionally introduces significant errors.

In another approach, discussed in Ref. [5], a table of standard chemical availabilities is developed based on reference substances selected to be $O_2(g)$, $N_2(g)$, $CO_2(g)$, $H_2O(l)$, $SO_2(g)$, Al_2O_3 (crystal), and Fe_2O_3 (crystal), each pure at 25°C and 1 atm (see Problem 7-18). Since this choice is not closely matched to the natural environment, use of the table is not recommended for evaluating the availability in effluent streams or for calculating second law efficiencies requiring availability *magnitudes*. It is suitable, however, for determining irreversibilities and availability *differences*.

Table 7-5

Standard Chemical Availability for Selected Elements (Ref. [6])

$T_o = 298.15$ K, $p_o = 1.01325$ bar

Element	Reference Substance	$\bar{a}^{ch}$ (kJ/kgmol)
Al (solid)	Al_2SiO_5 (solid)	887,890
Ar (gaseous)	Ar (gaseous)	11,690[a]
C (solid)	CO_2 (gaseous)	410,530
Ca (solid)	Ca^{++} (ion)	717,400
Cu (solid)	Cu^{++} (ion)	134,400
Fe (solid)	Fe_2O_3 (solid)	377,740
H_2 (gaseous)	H_2O (gaseous)	238,850
N_2 (gaseous)	N_2 (gaseous)	720[a]
O_2 (gaseous)	O_2 (gaseous)	3,970[a]
S (solid-rhombic)	SO_4^{--} (ion)	598,850

[a] $\bar{a}_k^{ch} = -\bar{R}T_o \ln(p_k^o/p_o)$. For Ar, $p_k^o = 9.07 \times 10^{-3}$ bar; for N_2, $p_k^o = 0.7583$ bar; for O_2, $p_k^o = 0.204$ bar.

7-7 Closure

This chapter concludes the formal presentation of the availability concept. The remainder of the book is devoted to applications of it. Accordingly, this is a good place to review the main ideas developed.

Availability is the maximum work obtainable from a *combined* system of control mass and environment as the control mass comes into thermal, mechanical, and chemical equilibrium with the environment. The maximum is obtained only when the process of the combined system is in every respect internally reversible. The availability magnitude depends on two states, the state of the control mass and that of the environment, and is a measure of the departure of the state of the control mass from that of the environment.

To show the relationship of the parts of this presentation to the whole the availability magnitude can be viewed as the sum of two contributions: the thermomechanical availability and the chemical availability. It is to be noted, though, that this distinction is not always convenient for the practical utilization of the availability concept.

The thermomechanical availability is the maximum work that can be done by the combined system as the control mass passes to the *restricted* dead state—that is, passes to a state in which it is in thermal and mechanical equilibrium with the environment. As shown in Sec. 3-4 the thermomechanical availability

$$= (E - U_o) + p_o(V - V_o) - T_o(S - S_o)$$

At the restricted dead state the composition of the control mass is not necessarily the same as that of the environment. In principle, a difference in composition can be exploited to obtain additional work from the combined system. The chemical availability is the maximum work achievable. There are two distinct aspects of the chemical availability idea considered in this book, depending on whether a particular substance C_k ($k = 1, n$) contained within the control mass is also a component of the environment *or* is absent from it.

If C_k is a component of the environment, its contribution to the chemical availability of the control mass, per mole of C_k, is determined in Sec. 6-3 to equal

$$(\mu_{ko} - \mu_k^o)$$

where μ_{ko} and μ_k^o denote, respectively, the chemical potential of C_k within the control mass at the restricted dead state and within the environment.

This chapter considers the case where C_k is not a component of the environment, but can be reacted with substances drawn from the environment to produce other environmental substances. Sec. 7-3 shows that the contribution of C_k to the chemical availability of the control mass is also given by the last expression, provided the chemical potential μ_k^o is the one associated with C_k *as if* it exists in the environment (see Eqs. (7-3b and c)).

Collecting results, the availability is

$$A = (E - U_o) + p_o(V - V_o) - T_o(S - S_o) + \sum_{k=1}^{n} N_k(\mu_{ko} - \mu_k^o) \qquad (a)$$

where N_k is the number of moles of substance C_k present in the control mass. The first underlined term is the thermomechanical availability and the second is the chemical availability.

The intermediate restricted dead state can be eliminated by use of Eq. (6-2b) to obtain a form which is often convenient for the practical utilization of the availability concept

$$A = E + p_o V - T_o S - \sum_{k=1}^{n} N_k \mu_k^o \qquad (b)$$

The companion flow availability expressions are readily obtained. Introducing Eq. (a), on a per mole of mixture basis, into $\bar{a}_f = \bar{a} + (p - p_o)\bar{v}$ (from Sec. 3-6) gives

$$\bar{a}_f = (\bar{h} - \bar{h}_o) - T_o(\bar{s} - \bar{s}_o) + \sum_{k=1}^{n} x_k(\mu_{ko} - \mu_k^o)$$

Similarly, with Eq. (b) on a per mole of mixture basis follows

$$\bar{a}_f = \bar{h} - T_o \bar{s} - \sum_{k=1}^{n} x_k \mu_k^o$$

The kinetic and potential energy terms are not shown in the last two expressions. The use of the last equation is illustrated in Secs. 7-5 and 8-3.

Selected References

1. AHRENDTS, J., "Reference States," *Energy*, **5,** August/September 1980, 667–677.
2. BRZUSTOWSKI, T. A., and P. J. GOLEM, "Second Law-Analysis of Energy Processes Part I: Exergy—An Introduction," *Trans. Can. Soc. Mech. Eng.*, **4,** 4, 1976–1977, 209–218.
3. GAGGIOLI, R. A., and P. J. PETIT, "Use the Second Law, First," *Chemtech*, **7,** August 1977, 496–506.
4. RIEKERT, L., "The Efficiency of Energy Utilization in Chemical Processes," *Chem. Eng. Sci.*, **29,** 1974, 1613–1620.
5. SUSSMAN, M. V., "Availability Analysis," *Energy Use Management, Proceedings of the International Conference*, October 1977, R. A. FAZZOLARE and C. B. SMITH, eds., II, Tucson, AZ, 1977 57–64.
6. SZARGUT, J., "International Progress in Second Law Analysis," *Energy*, **5,** August/September 1980, 709–718.
7. WEPFER, W. J., and R. A. GAGGIOLI, "Reference Datums for Available Energy," *Thermodynamics: Second Law Analysis*, American Chemical Society Symposium Series No. 122, R. A. Gaggioli, ed., American Chemical Society, Washington, DC, 1980.

Problems

7-1 Using property values from Appendix Table B-6 verify that the temperature T_p at the reactor exit for the problem of Example 7-1 is 3371°R.

7-2 Using Eqs. (7-3e) and (7-3f) evaluate the chemical availability of hydrogen, H_2, relative to the environment of Table 7-1. Repeat for the case where the gas phase of the environment is nearly saturated with water vapor: $x^o_{H_2O} = 0.0312$.

7-3 The accompanying table shows an environment consisting of a gas phase and a condensed water phase. The gas phase, which forms an ideal gas mixture, includes water vapor and also a number of dry components.

Environment $T_o = 298.15\ K\ (77\ °F),\ p_o = 1\ atm$	
Condensed phase: $H_2O(l)$ at T_o, p_o	
Gas phase: Substance	*Mole fraction* $(x°)$
N_2	0.7567
O_2	0.2035
H_2O	0.0312
CO_2	0.0003
other	0.0083

If hydrocarbon fuel C_aH_b, pure at T_o, p_o, reacts with oxygen from the environment to produce environmental components carbon dioxide and *liquid* water

according to
$$C_aH_b + (a + b/4)O_2 \longrightarrow aCO_2 + b/2\,H_2O(l)$$
show that the chemical availability of C_aH_b is given by,
$$\bar{a}_f^{ch}(T_o, p_o) = -\Delta G(T_o, p_o) + \bar{R}T_o \ln\left[\frac{(x_{O_2}^o)^{a+b/4}}{(x_{CO_2}^o)^a}\right]$$
where
$$\Delta G(T_o, p_o) = [a\,\bar{g}_{CO_2} + b/2\,\bar{g}_{H_2O(l)} - \bar{g}_f - (a + b/4)\bar{g}_{O_2}](T_o, p_o)$$

7-4 Using the result of Problem 7-3, evaluate the chemical availability of hydrogen. Compare with the values obtained in Problem 7-2. Discuss.

7-5 Showing all steps, develop Eq. (7-3g).

7-6 For the reactor of Fig. 7-4, verify by means of an energy equation that the temperature of the product gas is 1700°F.

7-7 Referring to Table 7-4, verify the availability values listed for the products.

7-8 Show that the chemical availabilities of the hydrocarbons listed in Table 7-2 are closely represented by the following expressions,
  · Gaseous hydrocarbons
$$\frac{\bar{a}^{ch}}{(\text{LHV})} = 1.033 + 0.0169(b/a) - \frac{0.0698}{a}$$
  · Liquid hydrocarbons
$$\frac{\bar{a}^{ch}}{(\text{LHV})} = 1.04224 + 0.011925(b/a) - \frac{0.042}{a}$$
where (LHV) is the lower heating value. The use of a computer is recommended.

7-9 Figure P7-9 shows a power plant with *back pressure heating*. The plant produces both power and warm water for heating. Neglecting pump work, a second law efficiency for the plant can be written as
$$\epsilon = \frac{\dot{W} + \dot{m}_w(a_{f6} - a_{f5})}{\dot{A}_f}$$
where $\dot{W}$ is the power produced, $\dot{m}_w$ is the mass flow rate of the heated water and $\dot{A}_f$ is the rate availability enters with fuel.

(a) Show that ϵ can be rewritten in the form
$$\epsilon = \epsilon_{sg}\left\{\frac{\eta_t(h_2 - h_a) + \epsilon_h[(h_3 - h_4) - T_o(s_3 - s_4)]}{(h_2 - h_1) - T_o(s_2 - s_1)}\right\}$$
where ϵ_{sg} is a second law efficiency for the steam generator
$$\epsilon_{sg} = \frac{\dot{m}_s(a_{f2} - a_{f1})}{\dot{A}_f}$$
η_t is the *isentropic turbine efficiency*, ϵ_h is a second law efficiency for the heater
$$\epsilon_h = \frac{\dot{m}_w(a_{f6} - a_{f5})}{\dot{m}_s(a_{f3} - a_{f4})}$$
and $\dot{m}_s$ is the flow rate of the cycle fluid.

(b) Evaluate ϵ for $T_o = 60°F$, $p_o = 1$ atm. See Fig. P7-9 for given data.

Problems

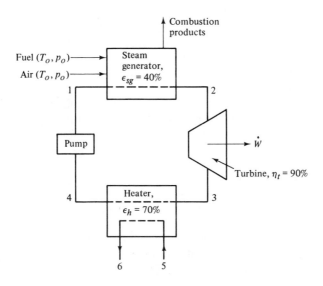

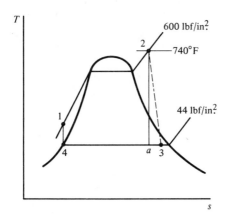

Figure P7-9

7-10 The following flow rates in lb/hr are reported for the exiting SNG (substitute natural gas) stream in a certain process for producing SNG from bituminous coal.

CH_4: 429,684 lb/hr
CO_2: 9,093 lb/hr
N_2: 3,741 lb/hr
H_2: 576 lb/hr
CO: 204 lb/hr
H_2O: 60 lb/hr
443,358 lb/hr

Assuming the SNG stream forms an ideal gas mixture and exits at 77°F, 1 atm, determine the rate availability exits. Perform calculations relative to the environment of Table 7-1. For further detail regarding this process, see S. P. Singh, S. A. Weil, and S. P. Babu, "Thermodynamic Analysis of Coal Gasification Processes," *Energy*, **5**, August/September 1980, 905–914.

7-11 Repeat Example 7-1 for the complete combustion of methane with the *theoretical* amount of air. Then calculate a second law efficiency for the reactor using Eq. (7-4a). Assume the fuel and air enter separately at temperature T_o and pressure p_o and the combustion products form an ideal gas mixture at pressure p_o. Perform availability calculations relative to the environment of Table 7-1.

7-12 Figure P7-12 shows a coal gasification reactor making use of the carbon-steam process. The energy required for the *endothermic* reaction is supplied by an electrical resistance heating unit. Assume the reactor operates at steady-state, with no stray heat transfers, and with negligible kinetic and potential energy effects.
  (a) Evaluate the required electrical energy input in Btu/lbmol of carbon.
  (b) Evaluate the availability entering with the carbon and with the steam in Btu/lbmol of carbon, relative to the environment of Table 7-1.
  (c) Evaluate the availability exiting with the product gas.
  (d) Determine the irreversibility for the reactor.

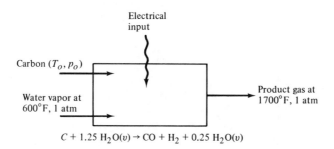

$$C + 1.25\, H_2O(v) \rightarrow CO + H_2 + 0.25\, H_2O(v)$$

Figure P7-12

7-13 Figure P7-13 shows a vapor power plant. The fuel is methane which is burned with air according to

$$CH_4 + 4(O_2 + 3.76 N_2) \longrightarrow CO_2 + 2H_2O(v) + 15.04 N_2 + 2O_2$$

Steam exits the steam generator at $T_1 = 900°F$, $p_1 = 500$ lbf/in.². The vapor expands through the turbine and exits at $p_2 = 1$ lbf/in.², $x_2 = 97\%$. At the condenser exit, $p_3 = 1$ lbf/in.² and the water is a saturated liquid. The pump work is negligible. Assume steady-state operation with no stray heat transfers from any plant component. Neglect all kinetic and potential energy effects. Base availability values on the environment of Table 7-1.
  (a) Determine the percent of the theoretical amount of air used in the reaction.
  (b) Determine the vapor mass flow rate per mole of fuel consumed.
  (c) Determine the cooling water mass flow rate per mole of fuel consumed.
  (d) Determine as a percent of the availability entering the steam generator with the fuel
    (i) the availability exiting with the stack gases
    (ii) the availability destroyed in the steam generator

Problems

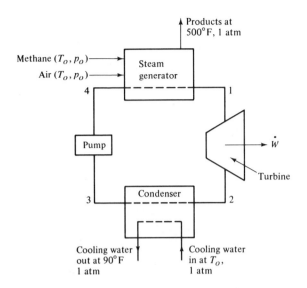

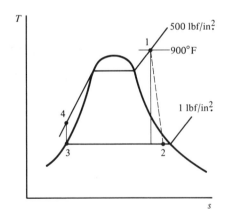

Figure P7-13

(iii) the power developed by the turbine
(iv) the availability destroyed in the turbine
(v) the availability exiting with the cooling water
(vi) the availability destroyed in the condenser

(e) Using the results of part (d) identify those plant components that offer the greatest opportunities for improvement through application of *practical engineering measures*. Discuss.

7-14 Consider a home furnace idealized as shown in Fig. P7-14. The fuel is methane which is burned with excess air according to

$$CH_4 + 4(O_2 + 3.76N_2) \longrightarrow CO_2 + 2H_2O(v) + 15.04N_2 + 2O_2$$

The furnace delivers a heat transfer $\bar{Q}_d$ at $T_d = 140°F$. The gaseous combustion products form an ideal gas mixture and exit at 620°F, 1 atm. There are no stray

heat transfers. Assume steady-state operation with negligible kinetic and potential energies.
(a) Evaluate $\bar{Q}_d$ in Btu/lbmol of fuel.
(b) Evaluate the availability of the fuel entering in Btu/lbmol of fuel, relative to the environment of Table 7-1.
(c) Evaluate the availability exiting with the gaseous combustion products and with the heat transfer $\bar{Q}_d$.
(d) Evaluate the irreversibility.
(e) Calculate a furnace (energy) efficiency in the form

$$\eta = \frac{\bar{Q}_d}{\text{(Fuel higher heating value)}}$$

(f) Calculate a second law efficiency in the form

$$\epsilon = \frac{\left(1 - \frac{T_o}{T_d}\right)\bar{Q}_d}{\bar{a}_f^{ch}}$$

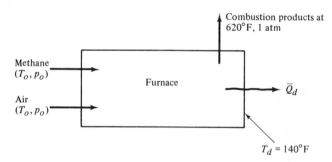

Figure P7-14

7-15 The accompanying table shows an environment consisting of a gas phase which forms an ideal gas mixture, a condensed water phase, and also solid gypsum, $CaSO_4 \cdot 2H_2O$, and solid calcite, $CaCO_3$. Determine the chemical availability relative to this environment for (a) CH_4, (b) SO_2, (c) COS, (d) H_2S, (e) NH_3.

Environment
$T_o = 298.15\ K\ (77\,°F),\ p_o = 1\ atm$

Condensed phase: $H_2O(l)$ at T_o, p_o
Solid phases: $CaSO_4 \cdot 2H_2O$, $CaCO_3$
Gas phase:

Substance	Mole fraction ($x°$)
N_2	0.7567
O_2	0.2035
H_2O	0.0312
CO_2	0.0003
other	0.0083

7-16 Table 7-5 gives the standard chemical availability for a number of elements as reported in Ref. [6]. With the aid of this table and Fig. P7-16, the standard chemical availability for each of several compounds can be calculated by means of the following procedure involving the reversible reaction of formation of the compound from the elements.

Assuming the reaction takes place at steady-state and at temperature T_o, an availability equation gives,

$$\bar{a}^{ch}_{compound} = -\bar{W}_{rev} + \sum_{k=1}^{n} \nu_k \bar{a}^{ch}_k$$

where $\bar{a}^{ch}_{compound}$ is the standard chemical availability of the compound of interest, $\bar{W}_{rev}$ is the shaft work developed in the reversible reaction per mole of compound, ν_k is the number of moles of the element k for each mole of compound, and $\bar{a}^{ch}_k$ is the standard chemical availability of element k (from Table 7-5).

By combining an energy equation with an entropy equation, $-\bar{W}_{rev} = \bar{g}_{compound}$, where $\bar{g}_{compound}$ is the Gibbs function of formation of the compound. So, finally

$$\bar{a}^{ch}_{compound} = \bar{g}_{compound} + \sum_k \nu_k \bar{a}^{ch}_k \qquad (a)$$

Showing all steps, develop Eq. (a).

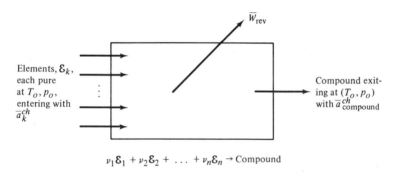

Figure P7-16

7-17 Following the procedure outlined in Problem 7-16, obtain the standard chemical availability for (a) CH_4, (b) SO_2, (c) COS, (d) H_2S, (e) NH_3.

7-18 In Ref. [5], the following is adopted as a reference for availability calculations. Each pure at $T_o = 25°C$ and $p_o = 1$ atm: $[O_2(g), N_2(g), CO_2(g), SO_2(g), H_2O(l), Al_2O_3(c), Fe_2O_3(c)]$ where g denotes a gas phase, l a liquid phase, and c a crystalline substance.

With the reference so defined, the standard chemical availability of a pure substance at 25°C, 1 atm is computed as its standard Gibbs function of formation minus the standard Gibbs functions of formation of its combustion products. For example, the compound $C_xH_yO_z$ has a standard chemical availability given by

$$\begin{pmatrix}\text{Standard chemical} \\ \text{availability of} \\ C_xH_yO_z \text{ at } T_o, p_o\end{pmatrix} = \bar{g}_{C_xH_yO_z}(T_o, p_o) - \left[x\,\bar{g}_{CO_2}(T_o, p_o) + \frac{y}{2}\bar{g}_{H_2O(l)}(T_o, p_o)\right] \qquad (a)$$

(a) Verify that Eq. (a) is a special case of Eq. (7-3a).
(b) Using the procedure described above, determine the standard chemical availability for (i) CH_4, (ii) SO_2, (iii) COS.

7-19 Compare and discuss the results obtained in Problems 7-15, 7-17, and 7-18, and the methods used to obtain them.

Chapter 8

Applications and Special Topics

The intent of this chapter is to show the breadth of application of the availability method of analysis and to introduce some special features of it which have not been considered in previous chapters.

8-1 Cryogenic Refrigeration System

Introduction. *Cryogenics* has been, and promises to continue to be, one of the most fruitful fields of application for availability analysis. Cryogenic systems achieve refrigeration at extremely low temperatures and typically have high power requirements and low efficiencies. The engineering literature contains many references to improved performance being realized as a result of investigations in which second law principles played a part. Their continued use in the low temperature field is assured by the emergence of technologies such as natural gas liquefaction and the cooling of long-distance electric power lines and superconductive electrical machinery.

To illustrate the application of availability principles in the field of cryogenics, a detailed analysis of a system using Helium 4 as the working fluid and providing refrigeration at 30 K is presented. The system, which closely resembles one suggested in Ref. [12], is shown schematically in Fig. 8-1a and in the accompanying temperature–entropy diagram of Fig. 8-1b. Steady-state operating data for the system are given in Table 8-1.

Summary. The results of a conventional energy analysis and an availability analysis are presented in Table 8-2.

A sharper picture of performance is provided by the results of the availabil-

8-1 Cryogenic Refrigeration System

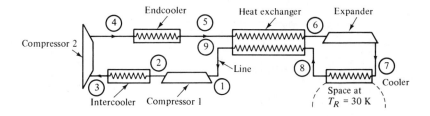

(a)

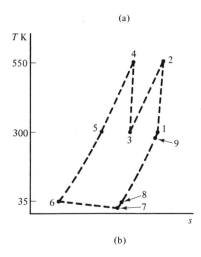

(b)

Figure 8-1 Helium refrigeration system.

Table 8-1

Operating Data for the Helium Refrigerator[a]

State	p (bars)	T (K)	Enthalpy (J/gmol)	Entropy (J/(gmol)(K))
1	1	300	6237	126.3
2	3.5	550	11437	128.4
3	3	300	6240	117.1
4	10.5	550	11446	119.2
5	10	300	6249	107.1
6	9	35	724	63.08
7	~1	19	393	68.81
8	~1	30	623	78.37
9	~1	296	6148	125.99

[a]Property values selected from S. Angus and K. M. de Reuck (Ed.), *International Thermodynamic Tables of the Fluid State Helium*-4, Pergamon Press, Elmsford, New York, 1977.

ity analysis. The availability analysis emphasizes that *both* losses and irreversibilities have an impact on system performance. The energy analysis focuses attention only on the energy losses, ignoring the irreversibilities. Furthermore,

Table 8-2

Summary of the Energy and Availability Analyses of the Helium Refrigerator

Component	Energy (J/gmol)			Availability (J/gmol) $T_o = 300$ K			
	Input	Output	Loss	Input	Output	Irreversibility, Loss of Availability[a]	
Compressor 1	5200	—	—	5200	—	630	(I)
Intercooler	—	—	5197	—	—	1807	(I + L)
Compressor 2	5206	—	—	5206	—	630	(I)
Endcooler	—	—	5197	—	—	1567	(I + L)
Heat Exchanger	—	—	—	—	—	1080	(I)
Expander	—	331	—	—	331	1719	(I)
Cooler	230	—	—	—	2070	568	(I)
Line	89	—	—	—	—	4	(I + L)
Subtotal	—	331	10394	—	2401	8005	
Total	10725		10725	10406		10406	

[a](I) denotes irreversibility, (I + L) irreversibility plus loss of availability with heat transfer.

the availability analysis shows the relative significance of the losses and irreversibilities in the various system components by evaluating them in terms of availability which takes into account both the *potential* and *quantity* of energy.

The remainder of this section is devoted to an explanation of the procedures followed in arriving at the values in Table 8-2.

Dead State. To perform the calculations underlying the values reported in Table 8-2, the only dead state property required is temperature T_o, taken to be 300 K. This is because irreversibilities and flow availability differences at fixed composition are the only calculations performed.

Adiabatic Units. For each of the adiabatic units (compressors, heat exchanger, expander) there is no loss of availability due to heat transfer, but there is an irreversibility which is found very readily by use of the relation $\bar{I} = T_o \bar{\sigma}$ where $\bar{\sigma}$ is the entropy production per mole for the unit determined from an entropy equation.

As an illustration, an entropy equation for the heat exchanger reduces to

$$0 = \bar{s}_5 + \bar{s}_8 - \bar{s}_6 - \bar{s}_9 + \bar{\sigma}$$

With entropy values from Table 8-1

$$\bar{\sigma} = 63.08 + 125.99 - 107.1 - 78.37$$
$$= 3.6 \text{ J/(gmol)(K)}$$

Multiplying by $T_o = 300$ K gives, $\bar{I} = 1080$ J/gmol.

8-1 Cryogenic Refrigeration System

Coolers. An availability equation for the intercooler reduces to

$$0 = \dot{N}_f(\bar{a}_{f2} - \bar{a}_{f3}) + \int_\alpha \left(1 - \frac{T_o}{T_s}\right) q_s \, d\alpha - \dot{I}$$

where $\dot{N}_f$ is the molar flow rate of the helium. Rearranging, and evaluating the flow availability difference

$$\frac{1}{\dot{N}_f}\left[\dot{I} - \int_\alpha \left(1 - \frac{T_o}{T_s}\right) q_s \, d\alpha\right] = (\bar{h}_2 - \bar{h}_3) - T_o(\bar{s}_2 - \bar{s}_3)$$

The left side of this equation represents the sum of the destruction of availability due to irreversibilities and the loss of availability associated with heat transfer from the unit. Evaluating the right side, the value in Table 8-2 is obtained. A similar calculation gives the value of Table 8-2 for the endcooler.

Consider next an availability equation for the cooler

$$0 = \dot{N}_f[\bar{h}_7 - \bar{h}_8 - T_o(\bar{s}_7 - \bar{s}_8)] + \left(1 - \frac{T_o}{T_R}\right)\dot{Q}_R - \dot{I} \qquad (a)$$

In Eq. (a), $\dot{Q}_R$ is the heat transfer to the unit from the refrigerated space at $T_R = 30$ K, and is found from an energy equation

$$\dot{Q}_R = \dot{N}_f(\bar{h}_8 - \bar{h}_7)$$
$$= \dot{N}_f(623 - 393) = \dot{N}_f(230)$$

Though the helium receives a heat transfer of energy from the refrigerated space, the associated flow of availability is *from* it *to* the refrigerated space. This follows because $T_R < T_o$.

$$\left(1 - \frac{T_o}{T_R}\right)\dot{Q}_R = \left(1 - \frac{300}{30}\right)(\dot{N}_f(230))$$
$$= -\dot{N}_f(2070).$$

Inserting the last result into Eq. (a), along with values from Table 8-1, gives $\bar{I} = 568$ J/gmol. Similar considerations show that although the line connecting ⑨ and ① (Fig. 8-1a) receives a heat transfer from the surroundings the associated flow of availability is out; it is counted as a loss.

Second Law Efficiency. A second law efficiency for the system in the form (product/net input) is

$$\epsilon = \frac{-\left(1 - \frac{T_o}{T_R}\right)\dot{Q}_R}{\text{Net input}}$$

$$= \frac{-\left(1 - \frac{T_o}{T_R}\right)\dot{Q}_R}{\dot{N}_f[(h_2 - h_1) + (h_4 - h_3) - (h_6 - h_7)]}$$

Inserting values

$$\epsilon = \left(\frac{2070}{5200 + 5206 - 331}\right) 100$$
$$= 20.5\%$$

This value should not be regarded as typical for devices in the class under consideration (see Problem 8-2).

In order to increase the efficiency it is necessary to find methods for reducing the irreversibilities and losses in those parts of the system where it is practical. For instance, there are well-known techniques for improving heat exchanger performance. Also, in large plants there may be economical uses for the *waste heat* of the coolers.

8-2 Turbojet Engine

Introduction. This section gives an availability analysis of a turbojet engine. The presentation is patterned after one given in Ref. [2].

A sketch of the engine is shown in Fig. 8-2. Steady-state operating data are listed in Table 8-3.

Table 8-3

Operating Data for the Turbojet Engine

Section	Fluid	$T(K)$	$p(bars)$	Velocity[a] (m/s)	Flow Rate (kg/s)
0	Air[b]	288	~1	0	1
1	Air	309.1	1.278	—	1
2	Air	548.8	7.667	—	1
—	Fuel (CH_4)	320	10	—	0.0257
3	Mixture[c]	1698.1	8.12	—	1.0257
4	Mixture	1448.7	4.275	—	1.0257
5	Mixture	985	~1	772	1.0257

[a]Magnitude of fluid velocity relative to the surrounding air at rest. See Sec. 3 of Ref. [2] for a discussion of velocity datums.
[b]For the air

	Flow Rate (kg/s)	Mole Fraction
N_2	0.78	0.79673
O_2	0.21	0.18771
CO_2	0.00035	0.00023
H_2O	0.00965	0.01533

[c]Mixture after complete combustion of CH_4 with air

	Flow Rate (kg/s)	Mole Fraction
N_2	0.78	0.7618
O_2	0.1072	0.0916
CO_2	0.071	0.0441
H_2O	0.0675	0.1025

8-2 Turbojet Engine

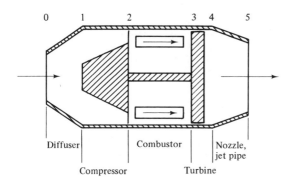

Figure 8-2 Turbojet engine.

Throughout the current presentation the fluids at work are assumed to be ideal gases with the constant specific heats listed in Table 8-4. This permits the use of a slightly different computational scheme than employed in any previous illustration.

Table 8-4

Specific Heat, c_p, and Gas Constant, R, for the Turbojet Analysis

	M	c_p (kJ/kg K)	$R = \bar{R}/M$ (kJ/kg K)
N_2	28	1.0437	0.29694
O_2	32	0.9329	0.25982
CO_2	44	0.8983	0.18896
H_2O	18	1.898	0.46191
CH_4	16	2.4153	0.51965
Air[a]	—	1.02862	0.29070
Mixture[a]	28.04	1.07826	0.29654

[a]Mixture values (see Sec. 2-4).

Summary. An availability analysis of the turbojet engine results in the values summarized in Table 8-5. The calculations are relative to the environment specified in the table.

The table shows that nearly 53% of the fuel's availability is carried out in the exhaust stream. The contributions associated with the relatively high exit temperature and exit velocity are about equal. Each component of the engine contributes to the overall irreversibility, with the combustor contribution, amounting to 29% of the fuel availability, being by far the dominant one. For the overall device, about 84% of the fuel availability is either destroyed by irreversibilities or carried out in the exhaust. The balance is available for *thrust power* (see Ref. [2] for a discussion).

In the remainder of this section the procedures used to determine the values of Table 8-5 are explained.

Table 8-5

Summary of the Availability Analysis for the Turbojet Engine[a]

	kJ/s	Percent of Fuel Availability
Availability in at 0[a]	0	—
Availability in with fuel	1347.71	—
Availability out, nozzle exhaust	713.38	52.93
Irreversibilities		
diffuser	0.41	0.03
compressor	20.06	1.49
combustor	388.58	28.83
turbine	5.61	0.42
nozzle, jet pipe	4.38	0.32
Total	419.04	31.09

[a]Calculations are relative to a gas phase consisting of still air at 288 K, 1 bar with a molar analysis as given for the air in note *b* of Table 8-3.

Inlet Air. Air enters the engine at temperature T_o and pressure p_o with the same composition as assumed for the environment. Thus, no availability enters with it. The only availability entering the engine is that associated with the fuel.

Fuel Availability. The fuel availability is calculated in this section using the procedure of Sec. 7-3.

The reaction of methane with environmental component oxygen to produce environmental components carbon dioxide and water vapor is

$$CH_4 + 2O_2 \longrightarrow CO_2 + 2H_2O(v)$$

The fuel chemical availability is

$$\bar{a}_f^{ch}(T_o, p_o) = [\bar{g}_f(T_o, p_o) + 2\bar{g}_{O_2}(T_o, p_o) - \bar{g}_{CO_2}(T_o, p_o) - 2\bar{g}_{H_2O(v)}(T_o, p_o)] \\ + \bar{R}T_o \ln\left[\left(\frac{x_{O_2}^o}{x_{H_2O(v)}^o}\right)^2 \frac{1}{x_{CO_2}^o}\right] \quad \text{(a)}$$

The Gibbs functions of Eq. (a) can be evaluated relative to values tabulated in Appendix B-7 at $T_{ref} = 298.15$ K, $p_{ref} = 1$ atm by use of Eqs. (7-2a and b). Thus, for the fuel

$$\bar{g}_f(T_o, p_o) = \bar{g}(T_{ref}, p_{ref}) + \{\bar{h}(T_o) - \bar{h}(T_{ref}) - (T_o\bar{s}(T_o, p_o) - T_{ref}\bar{s}(T_{ref}, p_{ref}))\}$$

Introducing the ideal gas model with constant specific heat $\bar{c}_p$, noting that $p_o \approx p_{ref}$, and rearranging

$$\bar{g}_f(T_o, p_o) = \bar{g}(T_{ref}, p_{ref}) + (\bar{c}_p - \bar{s}(T_{ref}, p_{ref}))(T_o - T_{ref}) - T_o\bar{c}_p \ln\left(\frac{T_o}{T_{ref}}\right)$$

where $\bar{s}(T_{ref}, p_{ref})$ is the absolute entropy for methane at 298.15 K, 1 atm. Inserting numerical values

8-2 Turbojet Engine

$$\bar{g}_f(T_o, p_o) = -50{,}846 + (38.64 - 186.27)(-10) - 288\left[38.64 \ln\left(\frac{288}{298}\right)\right]$$
$$= -48{,}991 \text{ kJ/kgmol}$$

Using the same procedure, the Gibbs functions for O_2, CO_2, and $H_2O(v)$, required by Eq. (a), are

$$\bar{g}_{O_2}(T_o, p_o) = 2046 \text{ kJ/kgmol}$$
$$\bar{g}_{CO_2}(T_o, p_o) = -392{,}545 \text{ kJ/kgmol}$$
$$\bar{g}_{H_2O(v)}(T_o, p_o) = -226{,}879 \text{ kJ/kgmol}$$

The natural logarithm contribution to Eq. (a)

$$= (8.315)(288) \ln\left[\left(\frac{0.18771}{0.01533}\right)^2 \left(\frac{1}{0.00023}\right)\right] = 32{,}060 \text{ kJ/kgmol}$$

Collecting results, the fuel chemical availability, $\bar{a}_f^{ch}(T_o, p_o)$ is 833,464 kJ/kgmol.

Since the fuel enters at $T_f = 320$ K, $p_f = 10$ bars, and not at T_o, p_o, the thermomechanical flow availability must be added to this sum. Omitting kinetic energy, the flow availability contribution

$$= (h(T_f) - h(T_o)) - T_o(\bar{s}(T_f, p_f) - \bar{s}(T_o, p_o))$$
$$= \bar{c}_{pf}(T_f - T_o - T_o \ln(T_f/T_o)) + \bar{R}T_o \ln(p_f/p_o)$$
$$= 5578 \text{ kJ/kgmol}$$

The availability entering with the fuel is then 839,042 kJ/kgmol, or 52,440 kJ/kg. With a fuel flow rate of 0.0257 kg/s, the value given in Table 8-5 is obtained.

Nozzle Exhaust. The availability leaving the control volume in the gas mixture at the nozzle exit is determined with Eq. (6-4f), modified to include kinetic energy

$$\dot{A}_s = \dot{N}_{mix} \sum_k x_k \left\{\bar{h}_k(T_s) - \bar{h}_k(T_o) - T_o(\bar{s}'_k(T_s) - \bar{s}'_k(T_o)) + \bar{R}T_o \ln\left(\frac{x_k}{x_k^o}\right)\right\} \quad \text{(b)}$$
$$+ \frac{\dot{m}_{mix} \mathcal{V}_s^2}{2}$$

where $\dot{N}_{mix}$, $\dot{m}_{mix}$ denote, respectively, the mixture flow rate on a molar and mass basis, $\dot{N}_{mix} = \dot{m}_{mix}/M_{mix}$. Introducing the mixture average specific heat $c_{p,mix}$, this reduces to

$$\dot{A}_s = \dot{m}_{mix} c_{p,mix} [T_s - T_o - T_o \ln(T_s/T_o)]$$
$$+ \dot{N}_{mix} \bar{R} T_o \sum_k x_k \ln\left(\frac{x_k}{x_k^o}\right) + \frac{\dot{m}_{mix} \mathcal{V}_s^2}{2} \quad \text{(c)}$$

Inserting values

$$\dot{A}_s = 379.18 + 28.55 + 305.65$$
$$= 713.38 \text{ kJ/s}$$

The terms associated with the elevated temperature and velocity of the exiting gas mixture contribute about equally to this sum.

Irreversibilities. The irreversibility for each component of the engine can be found easily by use of the relation $\dot{I} = T_o\dot{\sigma}$, where $\dot{\sigma}$ is the rate of entropy production determined from an entropy equation.

Considering the combustor first, an entropy equation gives

$$\dot{\sigma} = \sum_{\text{mix}} \dot{m}_i[s_i(T_{\text{ref}}, p_{\text{ref}}) + c_{pi} \ln(T_3/T_{\text{ref}}) - R_i \ln(x_i p_3/p_{\text{ref}})]$$
$$- \sum_{\text{air}} \dot{m}_i[s_i(T_{\text{ref}}, p_{\text{ref}}) + c_{pi} \ln(T_2/T_{\text{ref}}) - R_i \ln(x_i p_2/p_{\text{ref}})] \quad (d)$$
$$- \dot{m}_f[s_f(T_{\text{ref}}, p_{\text{ref}}) + c_{pf} \ln(T_f/T_{\text{ref}}) - R_f \ln(p_f/p_{\text{ref}})]$$

where the subscript f identifies the fuel and the terms $s(T_{\text{ref}}, p_{\text{ref}})$ are absolute entropy values on a unit mass basis. Inserting values and multiplying the result by T_o gives the combustor irreversibility reported in Table 8-5 (see Problem 8-5). (The total rate of entropy production from inlet to exit is given by Eq. (d) when T_3 and p_3 are replaced, respectively, by T_5 and p_5, and T_2, p_2 are replaced by T_o, p_o.)

The irreversibilities of the other components of the turbojet are also found through the calculation of the entropy production

$$\dot{\sigma}_{\text{diffuser}} = \dot{m}_{\text{air}}[c_{p,\text{air}} \ln(T_1/T_0) - R_{\text{air}} \ln(p_1/p_0)]$$
$$\dot{\sigma}_{\text{compressor}} = \dot{m}_{\text{air}}[c_{p,\text{air}} \ln(T_2/T_1) - R_{\text{air}} \ln(p_2/p_1)]$$
$$\dot{\sigma}_{\text{turbine}} = \dot{m}_{\text{mix}}[c_{p,\text{mix}} \ln(T_4/T_3) - R_{\text{mix}} \ln(p_4/p_3)]$$
$$\dot{\sigma}_{\text{nozzle}} = \dot{m}_{\text{mix}}[c_{p,\text{mix}} \ln(T_5/T_4) - R_{\text{mix}} \ln(p_5/p_4)]$$

8-3 Solid Waste Recovery System

Introduction. This section considers a solid waste recovery system that receives municipal waste from the surrounding community. The system, shown schematically in Fig. 8-3, consists of a processing plant and a power plant.

In the processing plant, the waste material trucked to the site is first passed through a shredder. Ferrous metal is removed with a magnetic belt separator and the remainder is sent to a second shredder. Next, the low density fraction, assumed to be the combustibles, is separated in an air density system and routed through a cyclone to remove dust and fines. The final product is transferred to a storage bin for eventual use as a supplemental boiler fuel in the power plant.

8-3 Solid Waste Recovery System

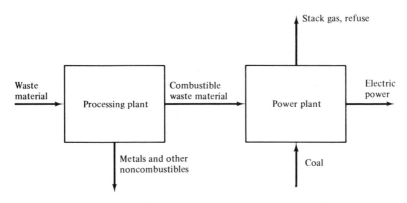

Figure 8-3 Schematic of solid waste recovery system.

The object of the current presentation is to evaluate the performance of the steam generator component of the power plant, shown in Fig. 8-4, for three cases, each at fixed boiler load. In one of these, the boiler is fired only with coal, in the other two cases different mixes of coal and waste material are used as fuel. An important feature of the presentation is that special means, not required in any previous illustration, are employed to evaluate the fuel availability.

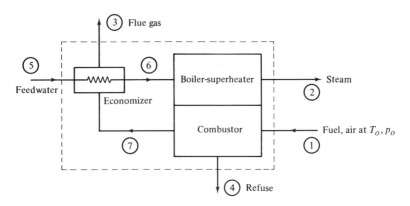

Figure 8-4 Schematic of steam generator.

Summary. Steady-state operating data for the three cases under consideration are given in Table 8-6. The results of the availability calculations performed are summarized in Table 8-7. This table also gives the environment used.

Second Law Efficiency. A second law efficiency for the steam generator can be written as the ratio of the availability increase for the water, as it passes from feedwater to steam, to the availability input with the fuel

$$\epsilon = \frac{a_2 - a_5}{a_1}$$

Table 8-6

Operating Data for the Solid Waste Recovery System (Ref. [9])

	Case I	Case II	Case III
Fuel flow rate (kg/hr)			
Coal	5487	3845 (62%)	2930 (43%)
Waste Material	0	2384 (38%)	3958 (57%)
Total	5487	6229	6888
Fuel analysis (kg/kg fuel)			
Carbon	0.498	0.422	0.384
Hydrogen	0.035	0.031	0.029
Oxygen	0.068	0.153	0.196
Sulfur	0.064	0.042	0.030
Free Moisture	0.141	0.167	0.181
Ash	0.194	0.185	0.180
Fuel heating value, $(LHV)_d$ (kJ/kg)			
	20,453	17,477	15,997
Air flow rate (kg/kg fuel)			
	16.724	11.964	11.141
Flue gas analysis (mole fractions)			
N_2	0.7690	0.7555	0.7488
O_2	0.1198	0.1094	0.1156
CO_2	0.0666	0.0772	0.0760
H_2O	0.0415	0.0555	0.0577
SO_2	0.0031	0.0024	0.0019
Flue gas flow rate (kg/kg fuel)			
	17.323	12.599	11.825
Refuse analysis (% by mass)			
Carbon	0.1053	0.0975	0.0729
Hydrogen	0.0064	0.0074	0.0074
Sulfur	0.0253	0.0360	0.0228
Ash	0.8629	0.8591	0.8968
Refuse mass (kg/kg fuel)			
	0.1988	0.1966	0.1746
Refuse heating value (kJ/kg)			
	4720	4680	3740
Other Data			
Steam Flow Rate (kg/hr)	34,370	34,370	34,370
p_2 (bars)	41.8	41.8	41.8
T_2 (K)	726	726	726
T_3 (K)	480	493	489
T_4 (K)	298	298	298
T_5 (K)	419	431	425
T_6 (K)	461	474	470
T_7 (K)	566	572	573

Table 8-7

Summary—Availability Analysis of the Steam Generator Component of the Solid Waste Recovery System at Fixed Boiler Load[a]

	Case I (Fuel Flow Rate = 5487 kg/hr, 100% Coal)		Case II (Fuel Flow Rate = 6229 kg/hr, 62% Coal)		Case III (Fuel Flow Rate = 6888 kg/hr, 43% Coal)	
	Availability (kJ/kg fuel)	% of Fuel Availability	Availability (kJ/kg fuel)	% of Fuel Availability	Availability (kJ/kg fuel)	% of Fuel Availability
Stream						
a_1	22,084	100	19,212	100	17,783	100
$(a_2 - a_5)$	7,456	33.8 [b]	6,484	33.7 [b]	5,903	33.2 [b]
a_3	1,709	7.7	1,307	6.8	1,139	6.4
a_4	1,013	4.6	1,008	5.2	709	4.
a_7	2,394	—	1,780	—	1,613	—
$(a_6 - a_5)$	369	—	356	—	328	—
Irreversibility						
Boiler-Combustor	11,590	52.5	10,296	53.6	9,886	55.6
Economizer	316	1.4	117	0.6	146	0.8
Total	11,906	53.9	10,413	54.2	10,032	56.4

[a] Calculations are relative to an environment at $T_o = 77°F$, $p_o = 1$ atm with components
  Condensed and solid phases at T_o, p_o
  (H_2O, $CaCO_3$, $CaSO_4 \cdot 2H_2O$)
  Gas Phase (mole fractions)
  (0.7561N_2, 0.2034O_2, 0.0312H_2O, 0.0003CO_2, 0.009 other)
[b] Second law efficiency = $(a_2 - a_5)/a_1$.

In writing this the availability exiting in the refuse and the flue gas are regarded as losses. The efficiencies calculated using the above equation for the three cases under consideration are shown enclosed by boxes in Table 8-7.

In each of the three cases the efficiency ϵ is in the 33 to 34% range. Based on thermodynamic considerations, the use of waste material as a fuel supplement does not appear to have an adverse effect on performance at fixed load. However, a comprehensive analysis of performance should also bring in considerations such as costs, material handling problems, corrosion, slagging effects, and so on.

The remainder of this section is devoted to an explanation of the procedures followed in arriving at the availability values reported in Table 8-7.

Water Stream. The increase in availability of the water as it passes through the overall unit is evaluated simply, using Appendix Table B-2 for property values and

$$a_2 - a_5 = \frac{\dot{m}_s}{\dot{m}_f}[(h_2 - h_5) - T_o(s_2 - s_5)]$$

where $\dot{m}_s$ and $\dot{m}_f$ are the steam and fuel flow rates, respectively. Similarly, for flow through the economizer, the availability increase is

$$a_6 - a_5 = \frac{\dot{m}_s}{\dot{m}_f}[(h_6 - h_5) - T_o(s_6 - s_5)]$$

With these equations the values given in Table 8-7 are readily computed.

Flue Gas Streams. Paralleling the development of Sec. 7-5, the availability per mole of flue gas

$$\begin{aligned}
&= \{x_{N_2}[\bar{h}(T) - \bar{h}(T_o) - T_o(\bar{s}'(T) - \bar{s}'(T_o)) + \bar{R}T_o \ln (x_{N_2}/x_{N_2}^o)]_{N_2} \\
&+ x_{O_2}[\bar{h}(T) - \bar{h}(T_o) - T_o(\bar{s}'(T) - \bar{s}'(T_o)) + \bar{R}T_o \ln (x_{O_2}/x_{O_2}^o)]_{O_2} \\
&+ x_{CO_2}[\bar{h}(T) - \bar{h}(T_o) - T_o(\bar{s}'(T) - \bar{s}'(T_o)) + \bar{R}T_o \ln (x_{CO_2}/x_{CO_2}^o)]_{CO_2} \quad (a)\\
&+ x_{H_2O}[\bar{h}(T) - \bar{h}(T_o) - T_o(\bar{s}'(T) - \bar{s}'(T_o)) + \bar{R}T_o \ln (x_{H_2O}/x_{H_2O}^o)]_{H_2O} \\
&+ x_{SO_2}[\bar{c}_p(T - T_o) - T_o\bar{c}_p \ln (T/T_o) + \bar{R}T_o \ln (x_{SO_2}) + \bar{a}^{ch}(T_o, p_o)]_{SO_2}\}
\end{aligned}$$

The last term of Eq. (a) is the contribution of the SO_2 in the flue gas. To evaluate it a constant specific heat has been used:[1] $\bar{c}_p = 43$ kJ/(kgmol)(K). Also, for SO_2, Ref. [8] gives $\bar{a}^{ch}(T_o, p_o) = 295{,}736$ kJ/kgmol, relative to the environment of the current calculations.

With Eq. (a) the availability values reported in Table 8-7 for the flue gas at both the inlet and exit to the economizer can be calculated (see Problem 8-7).

Air. Assuming the combustion air enters at temperature T_o and pressure p_o with the same composition as that listed for the gas phase in the environment (Table 8-7), no availability enters with it. Accordingly, the availability entering the combustor is just that associated with the fuel.

Fuel. In all previous illustrations the fuel has been assumed to be a pure substance. However, fuels encountered in practice usually are not pure substances but involve a number of components. This is the case for the application at hand. And although the methodology of Sec. 7-3 is applicable in principle, the chemical availabilities for such fuels cannot be determined in this way because their absolute entropies (Gibbs functions of formation) are not known. It is necessary then to resort to approximate methods.

As motivated by the discussion of Table 7-2 (Sec. 7-3), a reasonable approximation often can be had by taking the fuel chemical availability to equal the experimentally determined lower (or higher) heating value of the fuel. In another approach, equations presented in Ref. [8] can be used for estimating the chemical availabilities for certain categories of fuels, namely, (1) solid fuels

---

[1] R. C. Reid and T. K. Sherwood, *The Properties of Gases and Liquids*, 2nd ed., McGraw-Hill, New York, 1966, p. 195.

8-3 Solid Waste Recovery System

containing carbon, hydrogen, oxygen, and nitrogen, (2) solid fuels containing sulfur in addition to C, H, O, N, and (3) both liquid and gaseous fuels containing C, H, O, N. The availability values are relative to the environment adopted for the current analysis. These equations are derived from theoretical considerations along the lines of the methodology of Sec. 7-3 for fuels in the categories listed having *known* values for their absolute entropies. Accordingly, accuracy is assured with them only for the fuels used in their development. If a substance under consideration is not one of these, but is merely similar, the value obtained is best regarded as a *plausible approximation*.

Returning to the consideration of the solid waste recovery system, the equation reported in Ref. [8] applicable for estimating the availability of the fuel in each of the three cases is[2]

$$a^{ch}(T_o, p_o) = (LHV)_d \left[1.0438 + 0.0013 \frac{h}{c} + 0.1083 \frac{o}{c} + 0.0549 \frac{n}{c} \right]$$
$$+ 6740 s \frac{kJ}{kg} \quad \text{(b)}$$

where h/c, o/c, and n/c denote, respectively, the mass ratio of hydrogen to carbon, oxygen to carbon, and nitrogen to carbon. The mass fraction of sulfur is s, in kg per kg of fuel. The quantity $(LHV)_d$ is the lower heating value for the fuel with the free moisture removed, expressed per unit of moist fuel; this is the heating value given in Table 8-6.

A sample calculation shows the use of Eq. (b). For Case I, Table 8-6 provides

$$(LHV)_d = 20{,}453 \text{ kJ/kg}$$

$$\frac{h}{c} = \frac{0.035}{0.498} = 0.0703$$

$$\frac{o}{c} = \frac{0.068}{0.498} = 0.1365$$

$$\frac{n}{c} = 0$$

$$s = 0.064$$

Inserting these into Eq. (b) gives 22,084 kJ/kg, the value reported in Table 8-7 for a_1. For comparison, the higher heating value of the coal, as received, is 21,222 kJ/kg.

Refuse. As shown by the values of Table 8-6, some combustible matter remains in the refuse, and so availability exits with it. Since the refuse temperature is taken equal to T_o for simplicity, its availability is entirely chemical availability.

[2] Applies only when $\frac{o}{c} \leq 0.666$ (Ref. [8])

The same difficulty prevents the accurate evaluation of the refuse availability as for the fuel, that is, the absolute entropy is not known. Accordingly, an approximate method is required.

The values given in Table 8-7 for the refuse are calculated using Eq. (b). To illustrate, from Case I of Table 8-6

$$\text{Heating value} = 4720 \text{ kJ/kg}$$

$$\frac{h}{c} = \frac{0.0064}{0.1053} = 0.0608$$

$$\frac{o}{c} = \frac{n}{c} = 0$$

$$s = 0.0253$$

Inserting these values into Eq. (b) gives $a^{ch}(T_o, p_o) = 5097$ kJ per kg of refuse, or 1013 kJ per kg of fuel.

Irreversibilities. There are irreversibilities associated with both the economizer and boiler-combustor units, though the latter is far more significant. These may be evaluated by means of availability equations as follows.

For the economizer

$$I/\dot{m}_f = a_5 + a_7 - a_3 - a_6$$
$$= (a_7 - a_3) - (a_6 - a_5)$$

and for the boiler-combustor

$$I/\dot{m}_f = a_1 + a_6 - a_2 - a_4 - a_7$$
$$= a_1 - a_7 - a_4 - [(a_2 - a_5) - (a_6 - a_5)]$$

Miscellaneous heat losses are assumed to take place at temperature T_o.

8-4 Availability of Sunlight

Introduction. Though there are several potential areas of application, the engineering literature includes relatively few examples of availability analyses dealing explicitly with thermal radiation. Applications mentioned later include the study of biochemical processes taking place in plants exposed to sunlight and other studies related to the potential of solar radiation. Additional areas include the evaluation of solar and radiative heating systems, and the study of irreversibilities associated with heat transfer in industrial furnaces and combustors.

This section is limited to an introduction to the availability of thermal radiation. There are two parts to the presentation. In the first part, a simplified

8-4 Availability of Sunlight

model is developed which relates radiation and availability concepts. In the second part, the model is used as a point of departure to review some of the literature pertaining to the availability of sunlight.

Background. All bodies emit electromagnetic radiation by virtue of their temperature. The character of the radiation depends on how hot the body is and on the nature of its surface. Thermal radiation has the usual properties of electromagnetic waves. It has the velocity of light and can be reflected and absorbed. It carries energy and, when reflected or absorbed, exerts a *pressure*. It can be thought of conveniently as a *gas of photons*.

If radiation is trapped in a vessel with perfectly reflecting walls, the photons move in random directions and rebound elastically from the walls.[3] This is analogous to the case of an ordinary gas contained in a vessel with impervious walls. The gas of photons can be regarded as a *simple* system.

Using arguments drawn from the kinetic theory of gases, the energy of the photons contained in volume V is[4]

$$U = \alpha T^4 V \qquad (a)$$

where T is the temperature of the matter with which the radiation is in equilibrium and α is a constant. This is the well-known Stefan–Boltzman law. Furthermore, the radiation pressure resulting from photons impinging on the walls is

$$p = \frac{1}{3}\frac{U}{V}$$

$$= \frac{\alpha}{3}T^4 \qquad (b)$$

and the entropy associated with the gas is

$$S = \frac{4U}{3T} \qquad (c)$$

Availability. With the foregoing expressions for U, S, and p, all the ingredients are in hand for calculating the availability. As a first step, notice that with Eqs. (a) to (c), the term $(U_o + p_o V_o - T_o S_o)$ vanishes identically, and so

$$A = U - T_o S + p_o V$$

Using Eqs. (a) to (c) once again this becomes

$$A = U\left[1 - \frac{4}{3}\left(\frac{T_o}{T}\right) + \frac{1}{3}\left(\frac{T_o}{T}\right)^4\right] \qquad (8\text{-}4a)$$

---

[3] The remainder of this subsection and the next is limited to consideration of *isotropic* radiation.

[4] See for example, W. C. Reynolds, *Thermodynamics*, 2nd ed., McGraw-Hill, New York, 1968, Sec. 8.12.

(see Problem 8-10). To clarify the discussion of solar radiation given later in the section, the underlined term should be noted as the one associated with the radiation entropy.

Equation (8-4a) is used to prepare Table 8-8. The table values show that the term in brackets is always positive. Since it can be greater in value than unity, the occasional reference in the literature to it as an *efficiency* is misleading. The table also shows that $A = 0$ when $T = T_o$ and $A \rightarrow U$ as $T/T_o \rightarrow \infty$.

Table 8-8

$A/U = 1 - \frac{4}{3}[T_o/T] + \frac{1}{3}[T_o/T]^4$
(Eq. (8-4a))

T/T_o	A/U
0.5	3.67
0.63	1.
1.	0.
2.	0.35
5.	0.73
10.	0.87
20.	0.93
∞	1.

The Literature in Brief. The simple picture of isotropic radiation is not appropriate for sunlight. For one thing, *direct* sunlight has a directional character to which corresponds a lower entropy for the same energy than would be found for isotropic radiation. On the other hand, solar radiation reaching the earth is attenuated in the atmosphere through absorption and scattering. Since loss of directivity is an irreversible process, the potential of *diffuse* sunlight must be less than that for direct sunlight. With these ideas in mind this section concludes with a review of the contributions to this topic by a number of investigators. An important difference among these contributions is the way the term associated with the radiation entropy is developed. This term is shown underlined in each of the following three expressions.

In a discussion related to photosynthesis, radiation with energy U and temperature T_p is considered in Ref. [4]. The temperature T_p is that of solar radiation *after* scattering and is estimated to be 1350 K. By application of the energy and entropy equations, the maximum energy available for photosynthesis is reported to be

$$U\left[1 - \frac{4}{3}\left(\frac{T_o}{T_p}\right) + \frac{1}{3}\left(\frac{T_o}{T_p}\right)^4\right]$$

In this expression, which agrees in form with Eq. (8-4a), T_o is the temperature of the surroundings and is taken to be 300 K. Inserting temperature values, the term in brackets is about 0.70.

Selected References

The last expression is also obtained in Ref. [7] for the case of *direct* sunlight; however, the temperature T_p is replaced by the temperature T_s of solar radiation *before* losing its directional character. With $T_s = 5800$ K, the maximum fraction available is found to be 0.93. For *diffuse* sunlight, Ref. [7] reports

$$U\left[1 - \frac{4}{3}\left(\frac{T_o}{T'_s}\right)\right]$$

where $T'_s = T_s/\mathfrak{N}$ and $\mathfrak{N}$ is a number greater than one related to the solid angle occupied by the sun in the sky. The loss of directivity in diffuse sunlight serves to reduce the effective temperature, and thus the amount available. The numerical value of the term in brackets is given as 0.7.

Accounting for the fact that the direction of solar radiation is limited to a cone subtended by the sun's disc with half angle δ, the expression to follow is derived using availability principles in Ref. [5]

$$U\left[1 - \frac{4}{3}\left(\frac{T_o}{T_s^*}\right) + \frac{1}{3}\left(\frac{T_o}{T_s}\right)^4\right]$$

where $T_s^* = T_s/[1 - \cos(\delta)]^{1/4}$. Bringing in directionality in this way serves to increase the effective temperature, and thus the fraction available. With $\delta \approx 0.005$ radians, the term in brackets is about 0.996.

To summarize, there is general agreement among these references that the solar energy reaching the earth is to a considerable extent available for conversion to other uses, with direct sunlight being more so than diffuse. Further discussion of thermal radiation from the availability viewpoint is given in Refs. [3] and [6].

8-5 Closure

The applications considered in this chapter are representative of availability analyses which have appeared in the engineering literature. No attempt has been made in the current chapter, or in those leading up to it, to include an example of every type of system which has been, or might be, analyzed with the availability method. The cases considered have been selected to illustrate principles and bring out important points, while suggesting the breadth of application of the availability method of analysis.

Selected References

1. *ASHRAE Handbook and Product Directory*, "Fundamentals," American Society of Heating, Refrigerating, and Air-Conditioning Engineers, Inc., New York, 1977, Chap. 1.

2. CLARKE, J. M. and J. H., HORLOCK, "Availability and Propulsion," *J. Mech. Eng. Sci.*, **17**, *4*, 1975, 223–232.
3. EDGERTON, R. H., "Second Law and Radiation," *Energy*, **5**, August/September 1980, 693–707.
4. LANDSBERG, P. T., "A Note on the Thermodynamics of Energy Conversion in Plants," *Photochem. Photobiol.*, **26**, 1977, 313–314.
5. PARROTT, J. E., "Theoretical Upper Limit to the Conversion Efficiency of Solar Energy," *Solar Energy*, **21**, 1978, 227–229.
6. PETELA, R., "Exergy of Heat Radiation," *ASME J. Heat Trans.*, **86**, May 1964, 187–192.
7. PRESS, W. H., "Theoretical Maximum for Energy from Direct and Diffuse Sunlight," *Nature*, **264**, December 1976, 734–735.
8. RODRIQUEZ, L., "Calculation of Available-Energy Quantities," *Thermodynamics: Second Law Analysis*, American Chemical Society Symposium Series No. 122, R. A. Gaggioli, Ed., American Chemical Society, Washington, DC, 1980, 39–59.
9. SHAPIRO, H. N., and T. H., KUEHN, "Second Law Analysis of the Ames Solid Waste Recovery System," *Energy*, **5**, August/September 1980, 985–991.
10. SPANNER, D. C., *Introduction to Thermodynamics*, Academic, New York, 1964, 225–226.
11. SZARGUT, J., and T., STYRYLSKA, "Angenäherte Bestimmung der Exergie von Brennstoffen," *Brennstoff-Wärme-Kraft*, **16**, 1964, 589–596.
12. TREPP, C., "Refrigeration Systems for Temperatures Below 25 K with Turboexpanders," *Advances in Cryogenic Engineering*, **7**, 1961, 251–261.

Problems

8-1 Verify all entries in Table 8-2.

8-2 Referring to the helium refrigeration system of Sec. 8-1, let state 7 be defined by $T_7 = 21$ K, $h_7 = 435$ J/gmol, and $s_7 = 70.91$ J/(gmol)(K). Assuming the other states are as given in Table 8-1, determine the irreversibility of the expander and the second law efficiency for the overall refrigeration system.

8-3 Calculate the fuel availability value of Table 8-5, showing all steps.

8-4 Derive Eq. (c) (page 187) of Sec. 8-2 from Eq. (b) (page 187), and use it to calculate the availability exiting at the nozzle exit reported in Table 8-5. Show all details.

8-5 Verify the irreversibility values listed in Table 8-5.

8-6 Verify the values reported in Table 8-7 for the availability gain of the water stream as it passes through (1) the boiler-combustor, and (2) the economizer for (a) Case I, (b) Case II, and (c) Case III.

8-7 Verify the values reported in Table 8-7 for the flue gas availability at both the inlet and exit to the economizer for (a) Case I, (b) Case II, and (c) Case III.

8-8 Using Eq. (b) of Sec. 8-3 (page 193), determine the availability values reported in Table 8-7 for the fuel and the refuse for (a) Case II and (b) Case III.

8-9 Verify the irreversibility values reported in Table 8-7. Use the stream availability values given in the table.

8-10 Derive Eq. (8-4a) and verify the values of Table 8-8.

Chapter 9

Introduction to Thermoeconomics

Thermoeconomics combines principles drawn from the thermal sciences and the field of engineering economy for the purpose of rational decision making in the development and operation of effective energy systems. The role of the availability principle in thermoeconomic analyses is introduced in this chapter. A number of simplified illustrations are included to bring out important ideas.

9-1 Introduction

There are several factors involved when the question of how to improve the efficiency with which energy resources are utilized is taken up (see Sec. 1-3). It is not a purely technical matter. Many of these factors are economic at root, and the decision making process is usually dominated by cost considerations. Energy is but one contributor to total cost.

Thermoeconomics combines principles drawn from the thermal sciences (thermodynamics, heat transfer, and fluid mechanics) and the field of engineering economy for the purpose of rational decision making in the development and operation of effective energy systems. No attempt is made to examine this large subject in depth. The goal is to sketch the role of the availability concept in such analyses.

Though thermoeconomic analyses can be conducted without recognition of the availability concept, their formulation in many cases is more succinct and easily understood if it is used. This is because the availability concept takes into account both the quantity and potential (quality) of energy, and also permits an explicit and consistent accounting of both losses and irreversibilities. Accordingly, the use of availability for thermoeconomic analyses is featured in this chapter. Two areas of application are considered: costing and design. The aim is to present the fundamental principles while bringing in just enough complexity to suggest important practical considerations.

9-2 Engineering Economy Background

This section presents those basic engineering economy principles required to gain an understanding of the methodology of thermoeconomics and to equip the reader for further study of the literature. Most concepts are introduced briefly. Standard texts should be consulted if further elaboration is required on any point. Texts suitable for this purpose are listed at the end of the chapter.

Price and Cost. According to economic theory, in a *free* economy *price* is determined by the intersection of the supply and demand curves. The supply curve gives the relationship between the amount of an item available for sale and its price per unit. The demand curve gives the quantity people are willing to buy versus the unit price. However, factors such as cartel pricing and regulatory measures distort this simple picture.

The *cost* of an item is what is paid to acquire or produce it. In performing thermoeconomic analyses, it is costs that play the dominant role.

Cost Classifications. The *first cost* is the set of costs associated with the start of a project. As the name suggests, these costs normally occur only once for a given project.

The term *fixed cost* identifies those costs that remain relatively constant over a wide range of operational activity as measured by output or some other appropriate quantity. Costs related to taxes, insurance, interest, maintenance, administration, and so on are of this type.

Variable costs are those costs that vary more or less directly with the volume of output. These include the costs of materials, labor, fuel, and electrical power.

Using the concepts of fixed and variable costs, their relationship to the *total cost* is shown in Fig. 9-1. Note that a continuous variation is pictured. It is not uncommon, however, for costs to change in a stepped pattern with a change in operational activity.

The *unit cost* is the ratio of the total cost to the number of units. This is illustrated on Fig. 9-1 by the ratio Y/X. *Incremental* cost refers to the increase in cost for producing a specified number of additional units. Referring again to Fig. 9-1, y is the incremental cost of producing x additional units. When used in this chapter, the term *marginal* cost refers to the rate of change of cost with output. On Fig. 9-1, the marginal cost at p is the slope of the tangent at p.

In time, all devices become candidates for replacement due to obsolescence or some inadequacy. Normally, there are two alternatives. One is to retain the device currently in place for an additional time period. The other is to remove the current device and replace it with another.

9-2 Engineering Economy Background

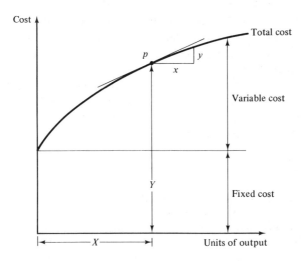

Figure 9-1 Fixed, variable and total cost versus output.

Replacement evaluations require the proper handling of *sunk* costs. A past cost or *sunk* cost is one that cannot be changed by future actions. Since only future consequences can be affected by current decisions, costs incurred in the past are ignored in replacement evaluations. Sunk costs are also disregarded in certain maintenance and operations decisions. A detailed example using the availability concept is presented in Ref. [4].

Asset Life. An asset is an item having monetary value. For engineering economy purposes, an asset can have several kinds of lives. Three of these are *ownership life*, *physical life*, and *economic life*.

The *ownership life* is the time period between the date of acquisition and date of disposal by a specific owner. The *physical life* of an asset is the time period from when it is new to when it is ultimately disposed of. Over its physical life an asset can have a succession of owners.

For any particular owner, the *economic life* of an asset is the time period dating from its installation to when it is removed from the intended *primary* service because the cost of a replacement asset is less than the cost of keeping it for an additional period. A strong factor in the replacement decision is often the increase in operating and maintenance costs which takes place as the asset provides service. At the end of an asset's economic life it is replaced but not necessarily disposed of, for it may pass into a new economic life for the same owner performing a *secondary* service role. If it is disposed of, it may go on to additional economic lives with a new owner.

The *salvage* value of an asset is the amount that is recovered or that could be recovered when it is removed from service.

Time Value of Money. *Interest* is the rental amount charged for the use of money. A dollar in hand today is worth more than a dollar received one year from now because having the dollar now allows the opportunity for investing it for the year. Money has *earning power*. It is this potential for growth that gives money its time value.

It can also be said that money has a time value because the *purchasing power* varies with time. For example, during periods of *inflation* the number of dollars required to obtain a given amount of goods increases over time. The methods to be considered in this chapter can be extended to account for the effect of inflation (see Problem 9-1). However, to simplify the presentation the concept of the time value of money is limited to the fact it has earning power.

The notion that money has a time value means that equal dollar amounts at different points in time have different worth. When alternative engineering options are evaluated, it is essential to account properly for the time value of money used both currently and in the future. This brings in the subject of life-cycle costing.

Life-Cycle Costing. The term life-cycle costing refers to the analysis of all costs associated with a device or facility over its *life*. This requires a systematic evaluation of all relevant costs: the first cost and the salvage value, operating costs for fuel, labor and material, costs for interest, insurance, depreciation, taxes, and so on.

A properly drawn life-cycle cost analysis requires an economic forecast of the future. It is necessary to make judgments about things like the general inflation rate and how interest rates and fuel costs will change. In predicting future costs, there are many uncertainties. Some future costs are potentially widely variable, being subject to a variety of economic pressures and political decisions. Also, the number of years selected as the life can significantly influence the outcome of the analysis and the conclusions drawn from it. Accordingly, while life-cycle costing analyses are quantitative in character, it should be recognized that the numerical values used for the required parameters are determined from some combination of forecasting, established corporate policy, experience and judgment, and so on. Before attempting a detailed life-cycle cost analysis it is recommended that engineers consult qualified experts as well as the literature on this subject.

It is often convenient to deal with life-cycle costs on an *annual* basis. The following discussion introduces this idea and also brings out the roles of the *present-worth factor* and the *capital-recovery factor* in *annualizing* costs.

Present-Worth Factor. Suppose $100 is borrowed at an interest rate of 10% per year. At the end of one year the interest is $10. The principal plus interest is $(100)(1 + 0.1) = \$110$. If the borrower does not pay the interest incurred at the end of each loan period and if interest is charged on the total amount owed (principal plus interest), the interest is said to be *compounded*.

9-2 Engineering Economy Background

At the end of the second year, the total owed is $(110)(1 + 0.1) = (100)(1 + 0.1)^2 = \$121$. And at the end of n years the total is $\$(100)(1 + 0.1)^n$.

From this example it can be seen that the value Y after n annual interest periods of an original sum of money P at an annual interest rate i is

$$Y = P(1 + i)^n$$

The initial sum of money appreciates by a factor $(1 + i)$ each year.

Conversely, a future sum Y has a *present worth* P

$$P = \frac{Y}{(1 + i)^n} \tag{a}$$

That is, the present worth of a given future sum Y diminishes by a factor $(1 + i)^{-1}$ for each year in the future.

The factor $(1 + i)^{-n}$ is called the *present-worth factor* PWF(i, n).

$$\text{PWF}(i, n) \equiv (1 + i)^{-n} \tag{b}$$

A dollar amount at a time n years in the future multiplied by the present-worth factor gives its present worth. The use of the present-worth factor is illustrated in Example 9-1.

Capital-Recovery Factor. Consider n amounts $(Y_1, Y_2, \ldots, Y_n)$, where Y_m is a dollar amount to be paid (or received) at the end of year m ($m = 1, n$). Using Eq. (a), the present worth of Y_m is

$$\frac{Y_m}{(1 + i)^m}$$

Forming the present worth of each of the n amounts Y_m, their sum is

$$S = \sum_{m=1}^{n} \frac{Y_m}{(1 + i)^m}$$

For the special case of *equal* annual amounts

$$S = Y \sum_{m=1}^{n} (1 + i)^{-m}$$

Using the expression for the sum of a geometric progression, the last expression becomes

$$S = Y \left[\frac{1 - (1 + i)^{-n}}{i} \right] \tag{c}$$

(see Problem 9-2). This is the present worth of a series of n equal annual amounts Y.

Alternatively, the series of n equal annual amounts Y whose present worth is S is

$$Y = S\left[\frac{i}{1-(1+i)^{-n}}\right]$$

The term in square brackets is called the *capital-recovery factor* CRF(i, n).

$$\text{CRF}(i, n) \equiv \frac{i}{1-(1+i)^{-n}} \qquad (d)$$

Table 9-1 shows how the capital-recovery factor varies with interest rate i and number of years n. The use of the capital-recovery factor is illustrated in Example 9-1.

Table 9-1

Capital-Recovery Factor, CRF $\equiv i/[1-(1+i)^{-n}]$

n	4%	8%	12%	20%
1	1.04000	1.08000	1.12000	1.20000
2	0.53020	0.56077	0.59170	0.65455
3	0.36035	0.38803	0.41635	0.47473
4	0.27549	0.30192	0.32923	0.38629
5	0.22463	0.25046	0.27741	0.33438
10	0.12329	0.14903	0.17698	0.23852
15	0.08994	0.11683	0.14682	0.21388
20	0.07358	0.10185	0.13388	0.20536
25	0.06401	0.09368	0.12750	0.20212
100	0.04081	0.08004	0.12000	0.20000

Annualized Costs. Some costs are incurred at or near the initiation of an activity and others at various times in the future. Certain future costs may be constant while others vary from year to year. Comparisons between alternative engineering options which accomplish the same goal are often based on annualized (levelized) costs.

When costs are annualized, each year of an asset's life is assigned the same cost figure. The sum of the annual costs is the total cost associated with the asset over its life taking into account the time value of money. The list below provides a simplified view of what can be a complex calculation.

- Letting Y_m be the cost associated with year m ($m = 1, n$), its present worth is determined by use of the present-worth factor.[1]
- The sum S of all present worth values is formed. This includes costs incurred at the initiation of the activity.

[1] For simplicity, assume Y_m is due at the end of year m.

9-2 Engineering Economy Background

- The capital-recovery factor is used to determine the series of n equal annual amounts whose present worth is S.

Application of this procedure leads to the following simple *cost equation*.

$$\dot{C}(\$/\text{year}) = \left\{[C_o - (SV)\text{PWF}(i, n)] + \sum_{m=1}^{n} Y_m \text{PWF}(i, m)\right\} \text{CRF}(i, n)$$

where C_o is the initial cost, (SV) denotes the salvage value at the end of year n, and Y_m represents the cost associated with operations for year m.

Example 9-1. Let the first cost of a certain device be \$10,000. The year to year operating costs Y_m are given in the table (the payment is assumed to take place at the end of the year indicated). For $i = 10\%$, determine the total present worth of all costs. Annualize the costs over the 5-year period. Let the salvage value after 5 years be zero.

Solution. To determine the present worth of the cost for year m requires use of the present-worth factor

$$\text{PWF}(i, m) = (1 + i)^{-m}$$

The result is given in the table below:

End of Year m	Cost, Y_m (\$)	PWF(10%, m)	Present Worth $= \text{PWF} \cdot Y_m$ (\$)
1	1000	0.9091	909.10
2	1500	0.8264	1239.60
3	2000	0.7513	1502.60
4	2000	0.6830	1366.00
5	2500	0.6209	1552.30
			\$6569.60

The total present worth S is the sum of the first cost, \$10,000, and \$6569.60: $S = \$16,569.60$.

The series of five equal annual payments whose present worth is S is determined by use of the capital-recovery factor.

$$\dot{C}\,(\$/\text{year}) = \text{CRF}\,(10\%, 5 \text{ years}) \cdot S$$
$$= (0.2638)(16{,}569.60)$$
$$= 4371.06$$

Costs are annualized by use of the capital-recovery factor which depends on the *life n* and the interest rate *i*. The values used for *n* and *i* determine the value of the CRF and through it can have a significant effect on how the item being evaluated is perceived. As shown by Table 9-1, at any specified value of *n*, the CRF increases with *i*. For fixed *i*, the CRF decreases as *n* increases. For very large values of *n* the CRF varies little with *n* and is determined mainly by *i* (see Problem 9-3). The role of the parameters *i* and *n* in annualizing costs is considered further in Example 9-2 and Problem 9-5.

9-3 Cost Equations

This section and the next consider the role of the availability concept in formulating cost equations for use in costing and design applications. A number of elementary illustrations are provided to help bring out the main ideas.

Introduction. Consider a device having a single fuel input which at steady-state produces a single output product in the form of electricity, shaft work, a desired heat transfer, etc. The cost of the fuel consumed over any time period is assumed to be a significant fraction of the total cost for that period. For such a device it is convenient to represent the annualized cost of producing the product as a sum of the fuel cost and costs for creating and maintaining the device.[2]

In symbols
$$(\$/year)_{product} = (\$/year)_{fuel} + (\$/year)_{nonfuel}$$

$$\dot{C}_p = \dot{C}_f + \dot{C}_{nf}$$

where $\dot{C}_f$ is the annualized cost of the fuel and $\dot{C}_{nf}$ is the annualized cost of all nonfuel items.

The costs $\dot{C}_f$ and $\dot{C}_p$ can be evaluated in terms of a number of measures. As examples, in the case of fuel oil $\dot{C}_f =$ (\$/gallon)(gallons/year), and for steam as the product $\dot{C}_p =$ (\$/lb)(lb/year). Another possibility is to use availability as a *common* measure by writing

$$\dot{C}_f = c_f \dot{A}_f$$
$$\dot{C}_p = c_p \dot{A}_p$$

where $\dot{A}_f$ is the rate of availability input with fuel (Btu/year) and c_f is the unit

---

[2] To focus on the main ideas related to the development and use of cost equations, the question of how much value ought to be assigned to fuel, labor, material, time (as reflected in interest rates), and so on is not taken up. It is assumed that all required input cost data are known and that means are available to take into account uncertainty about present costs and the possibility for changes in the future.

9-3 Cost Equations

cost of the fuel based on availability ($/Btu). The rate of availability output (Btu/year) is $\dot{A}_p$ and c_p is its unit cost based on availability ($/Btu).

Collecting the last three equations

$$c_p \dot{A}_p = c_f \dot{A}_f + \dot{C}_{nf}$$

and introducing a second law efficiency for the device in the form (output/input), $\epsilon = \dot{A}_p/\dot{A}_f$

$$c_p = \frac{c_f}{\epsilon} + \frac{\dot{C}_{nf}}{\dot{A}_p} \quad \text{(a)}$$

Equation (a) shows that the unit cost of the product is necessarily greater than the unit cost of the input fuel because (1) $\epsilon < 1$ (availability is destroyed due to irreversibilities within the device and there may also be losses) and (2) there is a cost associated with creating and maintaining the device. If steps are taken to increase the second law efficiency, the first term on the right side is reduced. However, the unit cost of the product is not necessarily reduced, because improvements made to increase ϵ might well cause an increase in nonfuel costs. The goal of minimizing total cost requires a tradeoff between fuel and nonfuel costs.

Example 9-2. Two motor-driven electrical generating units are being considered for use at remote job sites. For unit A the first cost is $C_o = \$10{,}000$ and $\epsilon = 10\%$. For unit B, $C_o = \$6000$, $\epsilon = 6\%$. Each unit is said to have a *life* of 20 years after which its salvage value is zero. To simplify further, neglect all other nonfuel costs. Compare the units for interest rates $i = 4\%, 8\%, 12\%, 15\%$, and 20%. In making the comparison, assume the same annual output $\dot{A}_p$ and unit fuel cost for each device, and let the product of these two known parameters be $c_f \dot{A}_p = 100$ ($/year). Repeat the comparison for a 50% increase in unit fuel cost.

Solution. The annualized nonfuel cost is $\dot{C}_{nf} = \text{CRF}(i, n) C_o$. Inserting this into Eq. (a) and rearranging gives

$$c_p = \left[\frac{1}{\epsilon} + \frac{\text{CRF}(i, n) C_o}{c_f \dot{A}_p}\right] c_f$$

The required comparison can be made through the ratio

$$\frac{(c_p)_A}{(c_p)_B} = \frac{\dfrac{1}{\epsilon_A} + \dfrac{\text{CRF}(i, n)(C_o)_A}{c_f \dot{A}_p}}{\dfrac{1}{\epsilon_B} + \dfrac{\text{CRF}(i, n)(C_o)_B}{c_f \dot{A}_p}}$$

Inserting values

$$\frac{(c_p)_A}{(c_p)_B} = \left[\frac{10 + 100\,\text{CRF}(i, 20)}{16.67 + 60\,\text{CRF}(i, 20)}\right]$$

Evaluating CRF(i, 20) for $i = 4\%$, 8%, 12%, 15%, and 20% gives the results in the following table:

i(%)	4	8	12	15	20
$\dfrac{(c_p)_A}{(c_p)_B}$	0.82	0.89	0.95	0.99	1.05

The table values show that the more fuel efficient unit A is preferred for interest rates up to 15%; but the lower first cost of unit B makes it the attractive option at interest rates higher than this.

At a fixed output $\dot{A}_p$, a 50% increase in unit fuel cost means that $c_f \dot{A}_p = 150$ (\$/year). Then

$$\frac{(c_p)_A}{(c_p)_B} = \left[\frac{10 + 66.67\,\text{CRF}(i, 20)}{16.67 + 40\,\text{CRF}(i, 20)}\right]$$

Substituting values for CRF(i, 20), it is easily verified that the more efficient unit A is preferred at each of the interest rates specified.

Costing.[3] For a device which produces a single output product using a single fuel input, the unit cost of the product can be determined using Eq. (a). When there is more than one output an important question concerns the way costs are distributed over the outputs. This is not a purely technical matter, but depends on the type of device, its technical purpose, and other economic considerations. And though availability is an index of economic value (Sec. 7-6), it is not appropriate in general to distribute costs proportionally to the availability rates of the outputs. These ideas are illustrated in the cases considered next.

In the first case, a steam turbine generator is considered. As shown in Fig. 9-2, steam enters at i and exits at e, and electricity is developed. A cost equation for the device is

$$c_e \dot{A}_e + (c_s \dot{A}_s)_e = (c_s \dot{A}_s)_i + \dot{C}_d \tag{b}$$

where $\dot{A}_s$ denotes a steam availability rate (Btu/year) and c_s is the corresponding

---

[3]Costing evaluations can require the use of *marginal* costs. There are cases of practical interest, however, where marginal costs are the same as unit costs, or approximately so. For an illustration see Eq. (9-4a) and the associated discussion.

9-3 Cost Equations

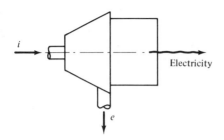

Figure 9-2 Steam turbine generating unit.

unit cost ($/Btu), c_e is the unit cost of electricity and $\dot{A}_e$ is the rate electricity (availability) exits. $\dot{C}_d$ accounts for the annual cost associated with creating and maintaining the device.

Regarding the right side of Eq. (b) and the rates $\dot{A}_e$ and $(\dot{A}_s)_e$ as known, the cost equation is a single relation in terms of two unknowns: the unit cost of electricity and the unit cost of the exiting steam. An additional relationship must be introduced for both unknowns to be determined. There are various methods for obtaining a second relationship (see Ref. [5]). One approach is to assign the *same* unit cost to each steam flow: $c_s \equiv (c_f)_i = (c_f)_e$. The result is that the primary output, electricity, is charged with the net steam availability used and the full cost of the device itself, as follows

$$c_e \dot{A}_e = c_s (\dot{A}_i - \dot{A}_e)_s + \dot{C}_d$$

where the first term on the right side accounts for the cost of the net steam availability used and the second for the device itself. In terms of the second law efficiency, $\epsilon_t = \dot{A}_e / (\dot{A}_i - \dot{A}_e)_s$, the unit cost of electricity is

$$c_e = \frac{c_s}{\epsilon_t} + \frac{\dot{C}_d}{\dot{A}_e} \tag{c}$$

Another control volume with more than one output is shown in Fig. 9-3. A cost equation for it is

$$c_2 \dot{A}_2 + c_3 \dot{A}_3 = c_1 \dot{A}_1 + \dot{C}_d$$

where $\dot{A}_i$ denotes an availability rate and c_i the corresponding unit cost. Let output stream 3 be an effluent stream. If it is simply discarded, it could be as-

Figure 9-3 Control volume.

signed a zero cost: $c_3 = 0$. The unit cost of the product at 2 is then easily determined. If the effluent stream must be treated before release to the surroundings, it could as a penalty be assigned a negative cost. This increases the unit cost of the product at 2, as can be seen from the cost equation.

Sample calculations illustrating the use of availability for costing are given in Ref. [5]. For an elementary case using Eq. (c) see Problem 9-6.

9-4 Design

The design process involves several steps, including identification of design requirements, formulation of alternative design configurations, and selection of parameters (dimensions, temperatures, flow rates, etc.) which define the best configuration. This section has to do with the third step, selection of the best configuration.

For the purposes of the present development, the best configuration is understood to be the one which minimizes total cost while satisfying all performance requirements. This is the meaning attached to the term *optimal*.

Illustration 1. To begin with a simple example, consider a device having a single fuel input which at steady-state produces a single output product. The annualized sum of costs for fuel and for creating and maintaining the device is the total annual cost, $\dot{C}_T$.

$$\dot{C}_T = \dot{C}_f + \dot{C}_{nf}$$

All other symbols have been previously defined.

Over the expected range of fuel use, let the fuel cost be described by

$$\dot{C}_f = c_f \dot{A}_f$$

where c_f is the unit cost of the fuel based on availability. Also, let the nonfuel costs vary with second law efficiency ϵ and output availability rate $\dot{A}_p$ as

$$\dot{C}_{nf} = K\epsilon \dot{A}_p$$

where K is a constant incorporating factors such as interest rate i and life n. The second law efficiency is $\epsilon = \dot{A}_p/\dot{A}_f$, where $\dot{A}_f$ is the rate availability enters with the fuel. It is recognized that devices are normally obtainable only in standard sizes with costs varying in fixed increments. The continuous variations given are assumed to approximate the actual situation and permit use of calculus.

Collecting these results, the annual cost of the product is

9-4 Design

$$\dot{C}_T = \left(\frac{1}{\epsilon} + \mathfrak{N}\epsilon\right) c_f \dot{A}_p$$

where $\mathfrak{N} = K/c_f$ is a dimensionless parameter which gauges the relative significance of nonfuel costs to fuel costs. For specified output and unit fuel cost, the sole *design variable* is the second law efficiency.

The optimal design is the one which minimizes the total cost. It can be determined in this case analytically by differentiating with respect to the efficiency and setting the result to zero, $d\dot{C}_T/d\epsilon = 0$. The optimal value for second law efficiency obtained in this way is $\epsilon = 1/\sqrt{\mathfrak{N}}$.

The illustrations to follow are intended to indicate that the process of optimization is generally more involved than suggested by this simple case.

Illustration 2. Figure 9-4 shows an insulated pipe. The figure is labeled with expressions determined in Sec. 3-7 for the loss of availability with heat transfer and the destruction of availability associated with fluid friction. As

$$\delta\dot{A}_{\text{loss}} = \frac{\left[1 - \dfrac{T_o}{T_f}\right]\left[T_f - T_\infty\right] \pi dL}{\left[\dfrac{1}{2k_p}\ln\left(\dfrac{D_2}{D_1}\right) + \dfrac{1}{2k_i}\ln\left(\dfrac{D_3}{D_2}\right) + \dfrac{1}{k_\infty Nu}\right]} \quad (3\text{-}7a)$$

$$\delta\dot{I} = \frac{T_o}{T_f}\left[\frac{8(\dot{m}_f)^3}{\pi^2 \rho^2 D_1^5}\right] f dL \quad (3\text{-}7b)$$

Figure 9-4 Insulated pipe.

discussed in Sec. 3-7, improved availability utilization can be brought about by reducing heat loss and fluid friction through the use of a smooth pipe with thick insulation and a relatively large diameter. <u>The extent to which these measures are adopted is limited, however, by the costs for insulation and piping which vary with size.</u> A rational selection of the pipe system necessarily must take these factors into consideration.

In the following discussion let the only unknowns be the pipe size and the insulation thickness. Assume fluid is delivered at a specified flow rate and exit condition, and all other parameters are known. The total annual cost is the sum of the cost of input availability and for the pipe system itself

$$\dot{C}_T = \dot{C}_i + \dot{C}_{\text{pipe}} \tag{a}$$

where $\dot{C}_{\text{pipe}}$ accounts for the cost of creating and maintaining the pipe. The cost of the availability entering is assumed to be described by

$$\dot{C}_i = c_i \dot{A}_i \tag{b}$$

where $\dot{A}_i$ is the rate availability enters and c_i is the unit cost based on availability.

The availability equation for the pipe is

$$\dot{A}_i = \dot{A}_e + \dot{A}_{\text{loss}} + \dot{I} \tag{c}$$

where $\dot{A}_{\text{loss}}$ is the loss of availability due to heat transfer and $\dot{I}$ accounts for the destruction of availability due to friction. Let $\dot{A}_{\text{loss}}$ and $\dot{I}$ be represented for simplicity as differentiable functions of the *design variables:* the nominal pipe diameter d and insulation thickness τ.

$$\dot{A}_{\text{loss}} = F(d, \tau)$$
$$\dot{I} = G(d, \tau)$$

Collecting Eqs. (a) to (c)

$$\dot{C}_T = c_i[\dot{A}_e + \dot{A}_{\text{loss}} + \dot{I}] + \dot{C}_{\text{pipe}}$$

The optimal design is the one which minimizes total cost. For c_i and $\dot{A}_e$ known, minimization of total cost is equivalent to minimization of

$$(\dot{C}_T - c_i \dot{A}_e) = c_i \dot{A}_{\text{loss}} + c_i \dot{I} + \dot{C}_{\text{pipe}} \tag{d}$$

All terms on the right side of Eq. (d) depend on the nominal pipe diameter and insulation thickness. The first term accounts for the annual cost associated with the loss of availability, the second term accounts for the annual cost of availability destruction associated with pipe friction, and the third term is the annualized cost of creating and maintaining the unit.

The minimization problem is approached, in principle, through the following necessary conditions

$$\frac{\partial(\dot{C}_T - c_i \dot{A}_e)}{\partial d} = 0$$

$$\frac{\partial(\dot{C}_T - c_i \dot{A}_e)}{\partial \tau} = 0$$

A computer solution offers the only practical approach, however, because of the complexity introduced by the friction factor f which depends on pipe diameter through both the Reynolds number and the relative pipe roughness and by the Nusselt number which depends on the diameter and insulation thickness. Additional complexity enters through the relationship of $\dot{C}_{\text{pipe}}$ to the design variables. In one numerical procedure, the nominal pipe diameter and insulation thickness are specified and the right side of Eq. (d) computed. The procedure continues for several additional choices for the pair (d, τ). The optimal selection is then found by searching the values computed. An example of this procedure is presented in Ref. [8].

Subsystem Analyses. It can be advantageous to break up an overall system into subsystems which, when individually optimized, optimize the overall system. One potential advantage of the subsystem approach is that there may be a reduction of mathematical complexity. For example, each subsystem might involve fewer independent variables than the overall system. Also, search procedures might be more efficiently implemented for a number of subsystems than for an overall system. Another potential advantage is that with relatively simple subsystems engineering insights can be brought to bear more effectively in seeking optimal designs.

The remainder of this section is directed toward bringing out some important considerations which enter when a system is decomposed into subsystems, and in particular the role of *marginal* costs at the junctures between subsystems.

Illustration 3. Figure 9-5 shows a system which can be reduced to two subsystems in series. On the figure, $\dot{A}_e$ is a known output availability rate. $\dot{A}_i$ is the input availability rate with unit cost c_i. $\dot{C}_a$ and $\dot{C}_b$ are the annualized costs of creating and maintaining units a and b, respectively. There are two design variables, X_a and X_b. X_a is associated with unit a and X_b is associated with unit b.

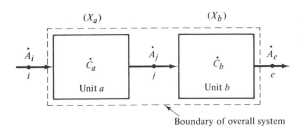

Figure 9-5 Two subsystems at steady-state in series with design variables X_a, X_b.

For each subsystem let the annual cost of creating and maintaining it, and the availability rate into it, be expressed as differentiable functions of the rate availability exits the subsystem and its *local* design variable which is X_a for unit a and X_b for unit b. That is, for unit a

$$\dot{A}_i = F(X_a, \dot{A}_j)$$
$$\dot{C}_a = f_a(X_a, \dot{A}_j) \tag{e}$$

where $\dot{A}_j$ denotes the availability rate at the juncture between the two units. For unit b, since $\dot{A}_e$ is known, the relations are simply

$$\dot{A}_j = G(X_b)$$
$$\dot{C}_b = f_b(X_b)$$

The cost equation for the overall system is

$$\dot{C}_T = c_i \dot{A}_i + \dot{C}_a + \dot{C}_b$$

For the cost to be a minimum, X_a and X_b are determined so that $\partial \dot{C}_T/\partial X_a = \partial \dot{C}_T/\partial X_b = 0$.

Alternatively, the total cost can be regarded to be a function of *three* variables: X_a, X_b, and $\dot{A}_j$, and minimized subject to the *constraint*, $G(X_b) - \dot{A}_j = 0$. This minimization problem is easily approached through use of the *method of Lagrange multipliers* (see Appendix A-2). The first step is to form

$$L = \dot{C}_T + \lambda(G - \dot{A}_j)$$

where λ is an *undetermined Lagrange* multiplier.

The minimum cost occurs when the partial derivatives of L with respect to X_a, X_b, and $\dot{A}_j$ vanish.

$$\frac{\partial L}{\partial X_a} = \frac{\partial}{\partial X_a}(c_i F + f_a) = 0 \tag{f}$$

$$\frac{\partial L}{\partial X_b} = \frac{d}{dX_b}(\lambda G + f_b) = 0 \tag{g}$$

$$\frac{\partial L}{\partial \dot{A}_j} = \frac{\partial}{\partial \dot{A}_j}(c_i F + f_a - \lambda \dot{A}_j) = 0 \tag{h}$$

Reducing Eq. (h) gives

$$\lambda = \frac{\partial}{\partial \dot{A}_j}(c_i F + f_a) \tag{i}$$

That is, the Lagrange multiplier λ has the significance of a *marginal* cost of availability at the juncture between the two subsystems; it is an *indicator* of how much unit a charges unit b for the intermediate availability stream. In general, the marginal cost of availability $\dot{A}_j$ is not the same as its unit cost.

Equations (f), (g), and (i) are three equations for the three unknowns, X_a, X_b, and λ. The first step in one numerical solution procedure is to assume a value for the marginal cost λ. Then, Eq. (g) can be used to obtain X_b. With X_b known, $\dot{A}_j$ is known and Eq. (f) can be used to obtain X_a. Finally, the assumed value for λ can be checked with Eq. (i). The procedure continues iteratively until the desired convergence is realized. As an alternative numerical procedure, X_a and X_b can be assumed and Eq. (i) used to determine λ. Equations (f) and (g) are then used to check all values.

The foregoing development shows that the overall system can be optimized by optimizing the two subsystems individually. This requires use of *local* cost equations for the two units: $\dot{C}_{Ta} = c_i \dot{A}_i + \dot{C}_a$ for unit a and $\dot{C}_{Tb} = \lambda \dot{A}_j + \dot{C}_b$ for unit b. A key aspect of this procedure is that the value assigned to the availability $\dot{A}_j$ at the juncture between the two subsystems is *in general* a marginal cost and not a unit cost.

There are cases, however, where marginal costs are the same as unit costs. As one such case of practical interest, consider a system for which $\dot{A}_i$ and $\dot{C}_a$ vary directly with the availability output rate $\dot{A}_j$ and the design variable X_a is the second law efficiency; that is, for which Eqs. (e) appear as

$$\dot{A}_i = \dot{A}_j / \epsilon_a$$
$$\dot{C}_a = \tilde{f}_a(\epsilon_a) \dot{A}_j \qquad \text{(j)}$$

With these, the marginal cost of availability at the juncture between the two devices, as given by Eq. (i), is (see Problem 9-9),

$$\lambda = \frac{c_i \dot{A}_i + \dot{C}_a}{\dot{A}_j} \qquad (9\text{-}4a)$$

which shows that for this case the marginal cost λ does equal the unit cost of availability at the juncture of the two subsystems.

The procedure of thermoeconomic analysis using the Lagrange method can be generalized to cases in which there are several design variables. It is in this kind of application that the power of the approach is felt. Detailed examples are found in Refs. [2] and [3].

9-5 Closure

This chapter has sketched the role of the availability concept in thermoeconomics. A number of simplified illustrations have been included to bring out the main ideas. Detailed applications can be found in several of the references listed at the end of the chapter and in the growing technical literature of this field.

Selected References

1. DeGarmo, E. P., J. R. Canada, and W. G. Sullivan, *Engineering Economy*, Macmillan, New York, 1979.
2. El-sayed, Y. M., and R. B. Evans, "Thermoeconomics and the Design of Heat Systems," *Trans. A.S.M.E. J. Engr. Power*, **92**, January 1970, 27–35.
3. Evans, R. B., G. L. Crellin, and M. Tribus, "Thermoeconomic Considerations of Sea Water Demineralization," *Principles of Desalination*, K. W. Spiegler, Ed., Academic, New York, 1966.
4. Fehring, T., and R. A. Gaggioli, "Economics of Feedwater Heater Replacement," *Trans. A.S.M.E. J. Engr. Power*, **99**, July 1977, 482–489.
5. Gaggioli, R. A., "Proper Evaluation and Pricing of Energy," *Energy Use Management, Proceedings of the International Conference*, October 1977, R. A. Fazzolare and C. B. Smith, Eds., **II**, Tucson, AZ, 1977, 31–43.
6. Kreider, J. F., *Medium and High Temperature Solar Processes*, Academic, New York, 1979, Chap. 7.
7. Thuesen, H. G., W. J. Fabrycky, and G. J. Thuesen, *Engineering Economy*, 5th Ed., Prentice-Hall, Englewood Cliffs, New Jersey, 1977.
8. Wepfer, W. J., R. A. Gaggioli, and E. F. Obert, "Economic Sizing of Steam Piping and Insulation," *Trans. A.S.M.E. J. Eng. Industry*, **101**, 4, November 1979, 427–433.

Problems

9-1 The effect of inflation on the future value of an amount P held today is to reduce the future value by a factor $(1 + \alpha)$ per year, where α is the annual inflation rate. Show that the value Y after n years, taking into account both interest and inflation is

$$Y = P\left(\frac{1+i}{1+\alpha}\right)^n$$

Use this to develop an *effective* interest rate i' which includes both interest and inflation effects

$$i' = \frac{i-\alpha}{1+\alpha}$$

The *present-worth* and *capital-recovery factors* can be modified to account for inflation by replacing the interest rate i with the effective interest rate i'.

9-2 Using the expression for the sum of a geometric progression, develop Eq. (c) of Sec. 9-2 (page 203).

9-3 Show that
(a) $\text{CRF}(i = 0, n) = 1/n$

Problems

    (b) $CRF(i, n = 1) = 1 + i$
    (c) $\lim_{n \to \infty} CRF(i, n) = i$

9-4 Let the initial cost of a certain device be $5000 and the operating cost in each year be $1000 (assume the payment is made at the end of the year). The salvage value at the end of 5 years is zero. For $i = 10\%$ annualize the costs over the 5-year period.

9-5 Repeat Example 9-2 for $n = 5$ years. Compare the results with those of Example 9-2. Discuss.

9-6 A steam turbine with effectiveness $\epsilon = 90\%$ develops shaft work equal to 10^8 kWh per year. The annual cost associated with creating and maintaining the turbine is 3×10^5 ($/year). The steam entering the turbine is valued at $.0165 per kWh of availability. Charging the shaft work developed with the net steam availability used and the full cost of the device itself, apply Eq. (c) of Sec. 9-3 (page 209) to determine its unit cost in $/kWh.

9-7 Consider a device which at steady-state produces a single output product using a single fuel input. Over the expected range of fuel use, let the fuel cost be described by: $\dot{C}_f = c_f \dot{A}_f$, where c_f is the unit cost of the fuel based on availability. Also, let the nonfuel costs vary with second law efficiency ϵ and output availability rate $\dot{A}_p$ as

$$\dot{C}_{nf} = k + K\left(\frac{\epsilon}{1-\epsilon}\right)\dot{A}_p$$

where k and K are constants incorporating factors such as interest rate i and life n, and $\epsilon = \dot{A}_p/\dot{A}_f$.

Show that for known output $\dot{A}_p$ and unit fuel cost c_f, minimum total cost is achieved when

$$\epsilon = \frac{1}{1 + \sqrt{K/c_f}}$$

9-8 Using the method of Lagrange multipliers, find minima and maxima for $z = xy$ subject to the constraint $x^2 + y^2 - 2 = 0$.

9-9 Verify Eq. (9-4a).

9-10 For the system shown in Fig. P9-10.

$$\dot{A}_i = \dot{A}_j/\epsilon_a$$
$$\dot{C}_a = \tilde{f}_a(\epsilon_a)$$

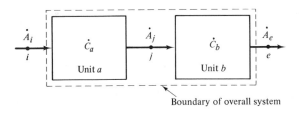

Boundary of overall system

Figure P9-10

That is, the design variable for unit a is the second law efficiency ϵ_a and the cost function $\dot{C}_a$ is independent of the exiting availability rate. Show that the *marginal* cost λ of availability at the juncture of the two subsystems is

$$\lambda = \frac{c_i}{\epsilon_a}$$

whereas its *unit* cost is

$$c_j = \frac{c_i}{\epsilon_a} + \frac{\tilde{f}_a(\epsilon_a)}{\dot{A}_j}$$

Appendix A

Reference Material

A-1 Unit Systems

The system of units used primarily in this book is the United States Customary System (USCS), although there is frequent use, particularly in Chaps. 7 and 8 of the International System (SI). Both systems adopt independent units for mass, length, time, and temperature. In the USCS, but not the SI, an independent unit for force is also specified. Units for other quantities are determined in terms of the independent units through definitions and physical laws. Table A-1 gives a partial listing of physical quantities and their units in both systems.

Table A-1

Unit Systems

Quantity	SI (International System) Unit	Symbol	USCS (United States Customary System) Unit	Symbol
Mass	kilogram	kg	pound mass	lb
Length	meter	m	foot	ft
Time	second	s	second	s
Temperature	degree Kelvin	K	degree Rankine	°R
Force	newton	N $(= 1 \text{ kg} \cdot \text{m/s}^2)$	pound force	lbf
Pressure	pascal	Pa $(= 1 \text{ N/m}^2)$	atmosphere	atm $(= 14.696 \text{ lbf/in.}^2)$
Energy	joule	J $(= 1 \text{ N} \cdot \text{m})$	foot-pound force	ft·lbf
Power	watt	W $(= 1 \text{ J/s})$	foot-pound force/s	ft·lbf/s

As it is often necessary to convert from one unit system to the other, a table of conversions is provided in Table A-2 to facilitate this process.

Table A-2

Selected Conversions[a]

Quantity	To Convert From	To	Multiply By
Mass			
	lb	kg	0.4536
	kg	lb	2.2046
Length, volume			
	ft	m	0.3048
	m	ft	3.2808
	in.^{3}	cm^3	16.387
	cm^3	in.^{3}	0.06102
	ft^3	m^3	0.02832
	m^3	ft^3	35.31
Force, pressure			
	N	lbf	0.22481
	lbf	N	4.4482
	lbf/in.^{2}	N/m^2	6894.757
	MPa	bars	10
	atm	lbf/in.^{2}	14.696
	atm	N/m^2	1.01325×10^5
	atm	bar	1.01325
Energy, power			
	J	N·m (W·s)	1
	kcal	kJ	4.1868
	Btu	kJ	1.0551
	Btu	kcal	0.252
	Btu	ft·lbf	778.17
	Btu/hr	kW	2.9307×10^{-4}
	Btu/hr	hp	3.93×10^{-4}
	hp	ft·lbf/s	550
	hp	kW	0.7457
Specific quantities			
	Btu/lb	kJ/kg	2.326
	Btu/lb°R	kJ/kg K	4.1868
	Btu/lb°R	kcal/kg K	1
	ft^3/lb	m^3/kg	0.062428

Additional Relations
  Temperature
   $T(°R) = 1.8 T(K)$
   $T(°C) = T(K) - 273.15$
   $T(°F) = T(°R) - 459.67$

  Universal Gas Constant

$$\bar{R} = \begin{cases} 8.315 \text{ kJ/(kgmol)(K)} \\ 1.986 \text{ kcal/(kgmol)(K)} \\ 1545 \text{ ft·lbf/(lbmol)(°R)} \\ 1.986 \text{ Btu/(lbmol)(°R)} \end{cases}$$

[a]Primary Source. ASTM E380–76. American Society of Testing and Materials, Philadelphia. The Btu is the International Table British thermal unit, the calorie is the International Table calorie. ASTM E380–76 is the same as IEEE 268–1976 and ANSI Z210.1–1976.

In the International System, the unit of force is defined from Newton's second law of motion in terms of the units for mass, length, and time, and is not taken as independent as in the USCS. That is, writing Newton's law as

$$F = ma$$

the unit of force, the newton (N), is defined to be the force required to accelerate a mass of one kilogram at the rate of one meter per second per second:

$$1 \text{ N} \equiv 1 \text{ kg} \cdot \text{m/s}^2$$

In the USCS, the unit of force, the pound force (lbf), is the force with which a standard pound mass is attracted to the earth at a location where the acceleration of gravity is 32.174 ft/s². Since the USCS makes use of *independent* units for force, mass, length, and time, Newton's second law of motion should be written as

$$F = \frac{ma}{g_c}$$

where g_c is a proportionality factor. Substituting the definition of the pound force, it follows that

$$g_c = 32.174 \frac{\text{lb} \cdot \text{ft}}{\text{lbf} \cdot \text{s}^2}$$

The constant g_c is often useful as a unit conversion factor. For example, if the linear kinetic energy, $m\mathcal{U}^2/2$, is being evaluated with mass in lb and the velocity in ft/s, its units are lb·(ft/s)². It can be converted to the unit ft·lbf by dividing by g_c.

$$\frac{m\mathcal{U}^2}{g_c} : \frac{\text{lb} \cdot (\text{ft/s})^2}{\left(\frac{\text{lb} \cdot \text{ft}}{\text{lbf} \cdot \text{s}^2}\right)} = \text{ft} \cdot \text{lbf}$$

In texts using the USCS exclusively, the factor g_c is often shown in the denominator of kinetic and potential energy terms. In this book, g_c is not shown in any equation. The reader should check the units of each term in an equation, and determine which require some type of conversion factor to get all quantities into a consistent set of units.

A-2 Method of Undetermined Lagrange Multipliers

A problem of considerable importance for many applications is that of maximizing or minimizing a function of several variables, where the variables are related by one or more equations. One approach is to use the equations to eliminate some of the variables and eventually reduce the problem to a conventional maximum or minimum problem. This is illustrated in the following example.

Example A-1. Minimize $z = x^2 + y^2$, where x and y are real numbers, subject to the constraining equation, $xy - 1 = 0$.

Solution. Solving the constraining equation, $y = 1/x$. With this

$$z = x^2 + \frac{1}{x^2}$$

Differentiating and setting the result to zero

$$\frac{dz}{dx} = 2x - \frac{2}{x^3} = 0$$

Solving this, the critical points are $(+1, +1)$ and $(-1, -1)$. Further consideration shows these are minimum points.

The solution procedure used in Example A-1 is not always convenient, or even feasible. Another approach is to use the *method of undetermined Lagrange multipliers*, stated formally as follows.

A *necessary condition* for a minimum (or maximum) of $F(x, y, \ldots)$ with respect to variables $x, y, \ldots$ that must satisfy the n constraining equations

$$C_i(x, y, \ldots) = 0 \qquad (i = 1, n) \tag{a}$$

is

$$\frac{\partial L}{\partial x} = \frac{\partial L}{\partial y} = \cdots = 0 \tag{b}$$

where

$$L \equiv F + \sum_{i=1}^{n} \lambda_i C_i$$

The constants $\lambda_1, \lambda_2, \ldots, \lambda_n$, called *undetermined Lagrange multipliers*, are evaluated, together with the minimizing (or maximizing) values of $x, y, \ldots$ by means of the set of equations consisting of (a) and (b). This is illustrated in Example A-2.

Example A-2. Repeat Example A-1 using the Lagrange method.

Solution. The first step is to form

$$L = (x^2 + y^2) + \lambda(xy - 1)$$

The necessary condition is $\partial L/\partial x = \partial L/\partial y = 0$. These two equations, along

A-2 Method of Undetermined Lagrange Multipliers

with the constraining equation $xy - 1 = 0$, give the system to be satisfied by x, y, and λ, namely

$$2x + \lambda y = 0$$
$$2y + \lambda x = 0$$
$$xy - 1 = 0$$

Solving this system, $\lambda = -2$ and the critical points are $(+1, +1)$ and $(-1, -1)$, as in Example A-1.

The method of Lagrange multipliers is used in Sec. 9-4 in the subsection headed *Illustration 3*.

Appendix B

Tables and Figures

B-1 Critical Properties

B-2 Properties of Water
   A. Saturated Water Temperature Table
   B. Saturated Water Pressure Table
   C. Superheated Vapor Table
   D. Compressed Liquid Table

B-3 Properties of Refrigerant 12(CCl_2F_2)
   A. Saturated Refrigerant 12 Temperature Table
   B. Saturated Refrigerant 12 Pressure Table
   C. Superheated Refrigerant 12 Table

B-4 Properties of Nitrogen
   A. Saturated Nitrogen Temperature Table
   B. Superheated Nitrogen Table

B-5 Ideal Gas Specific Heat $\bar{c}_p(T)$

B-6 Ideal Gas Properties of Selected Gases
   A. Air
   B. Diatomic Nitrogen
   C. Diatomic Oxygen
   D. Carbon Monoxide
   E. Carbon Dioxide
   F. Water
   G. Diatomic Hydrogen

B-7 Values of the Enthalpy of Formation, $\bar{h}_F$, Gibbs Function of Formation, $\bar{g}_F$, and Absolute Entropy at 25°C and 1 atm

App. B Tables and Figures

Figure B-1 Generalized Compressibility Chart
Z versus T_r, p_r, v'_r

Figure B-2 Generalized Enthalpy Departure Chart
$(h^* - h)/T_c$ versus T_r, p_r

Figure B-3 Generalized Entropy Departure Chart
$(s^* - s)$ versus T_r, p_r

Table B-1

Critical Properties[a]

Substance	Symbol	T_c K	p_c atm	v_c $cm^3/gmol$	Z_c —
Air	—	132.41	37.25	93.25	—
Helium	He	5.19	2.26	58	0.308
Hydrogen	H_2	33.24	12.797	65	0.304
Nitrogen	N_2	126.2	33.54	90	0.291
Oxygen	O_2	154.78	50.14	74	0.292
Carbon monoxide	CO	132.91	34.529	93	0.294
Carbon dioxide	CO_2	304.2	72.9	94	0.275
Sulfur dioxide	SO_2	430.7	77.8	122	0.269
Water	H_2O	647.27	218.167	56	0.230
Refrigerant 12	CCl_2F_2	385.16	40.6	217	0.279
Methane	CH_4	190.7	45.8	99	0.290
Ammonia	NH_3	405.4	111.3	72.5	0.243

[a]Data from references listed in Lydersen, A., R. Greenkorn, and O. Hougen, "Generalized Thermodynamic Properties of Pure Fluids," *Univ. Wisconsin Eng. Expt. Sta. Report 4*, October 1955.

Table B-2
Properties of Water[a]

Table B-2A
Properties of Saturated Water: Temperature Table v, ft^3/lb; h, Btu/lb; s, Btu/lb·°R

Temperature °F T	Pressure lbf/in.² p	Specific Volume		Internal Energy			Enthalpy			Entropy			Temp., °F T
		Sat. liquid v_f	Sat. vapor v_g	Sat. liquid u_f		Sat. vapor u_g	Sat. liquid h_f	h_{fg}	Sat. vapor h_g	Sat. liquid s_f	s_{fg}	Sat. vapor s_g	
32	0.0886	0.01602	3305	−0.01		1021.2	−0.01	1075.4	1075.4	−0.00003	2.1870	2.1870	32
35	0.0999	0.01602	2948	2.99		1022.2	3.00	1073.7	1076.7	0.00607	2.1704	2.1764	35
40	0.1217	0.01602	2445	8.02		1023.9	8.02	1070.9	1078.9	0.01617	2.1430	2.1592	40
50	0.1780	0.01602	1704	18.06		1027.2	18.06	1065.2	1083.3	0.03607	2.0899	2.1259	50
60	0.2563	0.01604	1207	28.08		1030.4	28.08	1059.6	1087.7	0.05555	2.0388	2.0943	60
70	0.3632	0.01605	867.7	38.09		1033.7	38.09	1054.0	1092.0	0.07463	1.9896	2.0642	70
80	0.5073	0.01607	632.8	48.08		1037.0	48.09	1048.3	1096.4	0.09332	1.9423	2.0356	80
90	0.6988	0.01610	467.7	58.07		1040.2	58.07	1042.7	1100.7	0.11165	1.8966	2.0083	90
100	0.9503	0.01613	350.0	68.04		1043.5	68.05	1037.0	1105.0	0.12963	1.8526	1.9822	100
110	1.276	0.01617	265.1	78.02		1046.7	78.02	1031.3	1109.3	0.14730	1.8101	1.9574	110
120	1.695	0.01621	203.0	87.99		1049.9	88.00	1025.5	1113.5	0.1647	1.7690	1.9336	120
130	2.225	0.01625	157.2	97.97		1053.0	97.98	1019.8	1117.8	0.1817	1.7292	1.9109	130
140	2.892	0.01629	122.9	107.95		1056.2	107.96	1014.0	1121.9	0.1985	1.6907	1.8892	140
150	3.722	0.01634	97.0	117.95		1059.3	117.96	1008.1	1126.1	0.2150	1.6533	1.8684	150
160	4.745	0.01640	77.2	127.94		1062.3	127.96	1002.2	1130.1	0.2313	1.6171	1.8484	160
170	5.996	0.01645	62.0	137.95		1065.4	137.97	996.2	1134.2	0.2473	1.5819	1.8293	170
180	7.515	0.01651	50.2	147.97		1068.3	147.99	990.2	1138.2	0.2631	1.5478	1.8109	180
190	9.343	0.01657	41.0	158.00		1071.3	158.03	984.1	1142.1	0.2787	1.5146	1.7932	190
200	11.529	0.01663	33.6	168.04		1074.2	168.07	977.9	1145.9	0.2940	1.4822	1.7762	200
210	14.13	0.01670	27.82	178.10		1077.0	178.14	971.6	1149.7	0.3091	1.4508	1.7599	210

[a] Abstracted from J.H. Keenan, F. G. Keyes, P. G. Hill, and J. G. Moore, "Steam Tables," John Wiley and Sons, New York, 1969. Reprinted by permission of John Wiley & Sons, Inc.

Table B-2A Continued

Temperature °F T	Pressure lbf/in.² p	Specific Volume		Internal Energy		Enthalpy			Entropy			Temp., °F T
		Sat. liquid v_f	Sat. vapor v_g	Sat. liquid u_f	Sat. vapor u_g	Sat. liquid h_f	h_{fg}	Sat. vapor h_g	Sat. liquid s_f	s_{fg}	Sat. vapor s_g	
212	14.70	0.01672	26.80	180.1	1077.6	180.2	970.3	1150.5	0.3121	1.4446	1.7567	212
220	17.19	0.01677	23.15	188.2	1079.8	188.2	965.3	1153.5	0.3241	1.4201	1.7441	220
230	20.78	0.01685	19.39	198.3	1082.6	198.3	958.8	1157.1	0.3388	1.3901	1.7289	230
240	24.97	0.01692	16.33	208.4	1085.3	208.4	952.3	1160.7	0.3534	1.3609	1.7143	240
250	29.82	0.01700	13.83	218.5	1087.9	218.6	945.6	1164.2	0.3677	1.3324	1.7001	250
260	35.42	0.01708	11.77	228.6	1090.5	228.8	938.8	1167.6	0.3819	1.3044	1.6864	260
270	41.85	0.01717	10.07	238.8	1093.0	239.0	932.0	1170.9	0.3960	1.2771	1.6731	270
280	49.18	0.01726	8.65	249.0	1095.4	249.2	924.9	1174.1	0.4099	1.2504	1.6602	280
290	57.53	0.01735	7.47	259.3	1097.7	259.4	917.8	1177.2	0.4236	1.2241	1.6477	290
300	66.98	0.01745	6.47	269.5	1100.0	269.7	910.4	1180.2	0.4372	1.1984	1.6356	300
310	77.64	0.01755	5.63	279.8	1102.1	280.1	903.0	1183.0	0.4507	1.1731	1.6238	310
320	89.60	0.01765	4.92	290.1	1104.2	290.4	895.3	1185.8	0.4640	1.1483	1.6123	320
330	103.00	0.01776	4.31	300.5	1106.2	300.8	887.5	1188.4	0.4772	1.1238	1.6010	330
340	117.93	0.01787	3.79	310.9	1108.0	311.3	879.5	1190.8	0.4903	1.0997	1.5901	340
350	134.53	0.01799	3.35	321.4	1109.8	321.8	871.3	1193.1	0.5033	1.0760	1.5793	350
360	152.92	0.01811	2.96	331.8	1111.4	332.4	862.9	1195.2	0.5162	1.0526	1.5688	360
370	173.23	0.01823	2.63	342.4	1112.9	343.0	854.2	1197.2	0.5289	1.0295	1.5585	370
380	195.60	0.01836	2.34	353.0	1114.3	353.6	845.4	1199.0	0.5416	1.0067	1.5483	380
390	220.2	0.01850	2.09	363.6	1115.6	364.3	836.2	1200.6	0.5542	0.9841	1.5383	390
400	247.1	0.01864	1.87	374.3	1116.6	375.1	826.8	1202.0	0.5667	0.9617	1.5284	400
420	308.5	0.01894	1.50	395.8	1118.3	396.9	807.2	1204.1	0.5915	0.9175	1.5091	420
440	381.2	0.01926	1.22	417.6	1119.3	419.0	786.3	1205.3	0.6161	0.8740	1.4900	440
460	466.3	0.01961	1.00	439.7	1119.6	441.4	764.1	1205.5	0.6404	0.8308	1.4712	460
480	565.5	0.02000	0.82	462.2	1118.9	464.3	740.3	1204.6	0.6646	0.7878	1.4524	480
500	680.0	0.02043	0.68	485.1	1117.4	487.7	714.8	1202.5	0.6888	0.7448	1.4335	500
540	961.5	0.02145	0.47	532.6	1111.0	536.4	657.5	1193.8	0.7374	0.6576	1.3950	540
580	1324.3	0.02278	0.32	583.1	1098.9	588.6	589.3	1178.0	0.7872	0.5668	1.3540	580
600	1541.0	0.02363	0.27	609.9	1090.0	616.7	549.7	1166.4	0.8130	0.5187	1.3317	600
640	2057.1	0.02593	0.18	668.7	1063.2	678.6	453.4	1131.9	0.8681	0.4122	1.2803	640
680	2705	0.03032	0.11	741.7	1011.0	756.9	309.8	1066.7	0.9350	0.2718	1.2068	680
700	3090	0.03666	0.07	801.7	947.7	822.7	167.5	990.2	0.9902	0.1444	1.1346	700
705.4	3204	0.05053	0.05	872.6	872.6	902.5	0	902.5	1.0580	0	1.0580	705.4

227

Table B-2B

Properties of Saturated Water: Pressure Table v, ft³/lb; h, Btu/lb; s, Btu/lb·°R

Absolute Pressure lbf/in.² p	Temperature °F T	Specific Volume Sat. liquid v_f	Specific Volume Sat. vapor v_g	Internal Energy Sat. liquid u_f	Internal Energy Sat. vapor u_g	Enthalpy Sat. liquid h_f	Enthalpy h_{fg}	Enthalpy Sat. vapor h_g	Entropy Sat. liquid s_f	Entropy s_{fg}	Entropy Sat. vapor s_g	Absolute Pressure lbf/in.² p
0.4	72.84	0.01606	792.0	40.94	1034.7	40.94	1052.3	1093.3	0.0800	1.9760	2.0559	0.4
1	101.70	0.01614	333.6	69.74	1044.0	69.74	1036.0	1105.8	0.1327	1.8453	1.9779	1
2	126.04	0.01623	173.75	94.02	1051.8	94.02	1022.1	1116.1	0.1750	1.7448	1.9198	2
4	152.93	0.01636	90.64	120.88	1060.2	120.89	1006.4	1127.3	0.2198	1.6426	1.8624	4
6	170.03	0.01645	61.98	137.98	1065.4	138.00	996.2	1134.2	0.2474	1.5819	1.8292	6
8	182.84	0.01653	47.35	150.81	1069.2	150.84	988.4	1139.3	0.2675	1.5383	1.8058	8
10	193.19	0.01659	38.42	161.20	1072.2	161.23	982.1	1143.3	0.2836	1.5041	1.7877	10
14.696	211.99	0.01672	26.80	180.10	1077.6	180.15	970.4	1150.5	0.3121	1.4446	1.7567	14.696
15	213.03	0.01672	26.29	181.14	1077.9	181.19	969.7	1150.9	0.3137	1.4414	1.7551	15
20	227.96	0.01683	20.09	196.19	1082.0	196.26	960.1	1156.4	0.3358	1.3962	1.7320	20
30	250.34	0.01700	13.75	218.84	1088.0	218.93	945.4	1164.3	0.3682	1.3314	1.6996	30
40	267.26	0.01715	10.50	236.03	1092.3	236.16	933.8	1170.0	0.3921	1.2845	1.6767	40
50	281.03	0.01727	8.52	250.08	1095.6	250.24	924.2	1174.4	0.4113	1.2476	1.6589	50
60	292.73	0.01738	7.18	262.1	1098.3	262.3	915.8	1178.0	0.4273	1.2170	1.6444	60
70	302.96	0.01748	6.21	272.6	1100.6	272.8	908.3	1181.0	0.4412	1.1909	1.6321	70
80	312.07	0.01757	5.47	282.0	1102.6	282.2	901.4	1183.6	0.4534	1.1679	1.6214	80
90	320.31	0.01766	4.90	290.5	1104.3	290.8	895.1	1185.9	0.4644	1.1475	1.6119	90
100	327.86	0.01774	4.43	298.3	1105.8	298.6	889.2	1187.8	0.4744	1.1290	1.6034	100
120	341.30	0.01789	3.73	312.3	1108.3	312.7	878.5	1191.1	0.4920	1.0966	1.5886	120
140	353.08	0.01802	3.22	324.6	1110.3	325.1	868.7	1193.8	0.5073	1.0688	1.5761	140

Table B-2B Continued

Absolute Pressure lbf/in.^{2} p	Temperature °F T	Specific Volume		Internal Energy		Enthalpy			Entropy			Absolute Pressure lbf/in.^{2} p
		Sat. liquid v_f	Sat. vapor v_g	Sat. liquid u_f	Sat. vapor u_g	Sat. liquid h_f	h_{fg}	Sat. vapor h_g	Sat. liquid s_f	s_{fg}	Sat. vapor s_g	
160	363.60	0.01815	2.84	335.6	1112.0	336.2	859.8	1196.0	0.5208	1.0443	1.5651	160
180	373.13	0.01827	2.53	345.7	1113.4	346.3	851.5	1197.8	0.5329	1.0223	1.5553	180
200	381.86	0.01839	2.29	354.9	1114.6	355.6	843.7	1199.3	0.5440	1.0025	1.5464	200
250	401.04	0.01865	1.84	375.4	1116.7	376.2	825.8	1202.1	0.5680	0.9594	1.5274	250
300	417.43	0.01890	1.54	393.0	1118.2	394.1	809.8	1203.9	0.5883	0.9232	1.5115	300
350	431.82	0.01912	1.33	408.7	1119.0	409.9	795.0	1204.9	0.6060	0.8917	1.4978	350
400	444.70	0.01934	1.16	422.8	1119.5	424.2	781.2	1205.5	0.6218	0.8638	1.4856	400
450	456.39	0.01955	1.03	435.7	1119.6	437.4	768.2	1205.6	0.6360	0.8385	1.4746	450
500	467.13	0.01975	0.93	447.7	1119.4	449.5	755.8	1205.3	0.6490	0.8154	1.4645	500
550	477.07	0.01994	0.84	458.9	1119.1	460.9	743.9	1204.8	0.6611	0.7941	1.4551	550
600	486.33	0.02013	0.77	469.4	1118.6	471.7	732.4	1204.1	0.6723	0.7742	1.4464	600
700	503.23	0.02051	0.66	488.9	1117.0	491.5	710.5	1202.0	0.6927	0.7378	1.4305	700
800	518.36	0.02087	0.57	506.6	1115.0	509.7	689.6	1199.3	0.7110	0.7050	1.4160	800
900	532.12	0.02123	0.50	523.0	1112.6	526.6	669.5	1196.0	0.7277	0.6750	1.4027	900
1000	544.75	0.02159	0.45	538.4	1109.9	542.4	650.0	1192.4	0.7432	0.6471	1.3903	1000
1200	567.37	0.02232	0.36	566.7	1103.5	571.7	612.3	1183.9	0.7712	0.5961	1.3673	1200
1400	587.25	0.02307	0.30	592.7	1096.0	598.6	575.5	1174.1	0.7964	0.5497	1.3461	1400
1600	605.06	0.02386	0.26	616.9	1087.4	624.0	538.9	1162.9	0.8196	0.5062	1.3258	1600
1800	621.21	0.02472	0.22	640.0	1077.7	648.3	502.1	1150.4	0.8414	0.4645	1.3060	1800
2000	636.00	0.02565	0.19	662.4	1066.6	671.9	464.4	1136.3	0.8623	0.4238	1.2861	2000
2500	668.31	0.02860	0.13	717.7	1031.0	730.9	360.5	1091.4	0.9131	0.3196	1.2327	2500
3000	695.52	0.03431	0.08	783.4	968.8	802.5	213.0	1015.5	0.9732	0.1843	1.1575	3000
3203.6	705.44	0.05053	0.05	872.6	872.6	902.5	0	902.5	1.0580	0	1.0580	3203.6

Table B-2C

Properties of Water: Superheated Vapor Table v, ft^3/lb; h, Btu/lb; s, Btu/lb°R

Temp., °F	v	u	h	s	v	u	h	s
	\multicolumn{4}{c}{10 psia (193.2°F)}	\multicolumn{4}{c}{14.7 psia (212.0°F)}
Sat.	38.42	1072.2	1143.3	1.7877	26.80	1077.6	1150.5	1.7567
200	38.85	1074.7	1146.6	1.7927				
250	41.93	1092.6	1170.2	1.8272	28.42	1091.5	1168.8	1.7832
300	44.99	1110.4	1193.7	1.8592	30.52	1109.6	1192.6	1.8157
400	51.03	1146.1	1240.5	1.9171	34.67	1145.6	1239.9	1.8741
500	57.04	1182.2	1287.7	1.9690	38.77	1181.8	1287.3	1.9263
600	63.03	1218.9	1335.5	2.0164	42.86	1218.6	1335.2	1.9737
700	69.01	1256.3	1384.0	2.0601	46.93	1256.1	1383.8	2.0175
800	74.98	1294.6	1433.3	2.1009	51.00	1294.4	1433.1	2.0584
900	80.95	1333.7	1483.5	2.1393	55.07	1333.6	1483.4	2.0967
1000	86.91	1373.8	1534.6	2.1755	59.13	1373.7	1534.5	2.1330
1100	92.88	1414.7	1586.6	2.2099	63.19	1414.6	1586.4	2.1674
	\multicolumn{4}{c}{20 psia (228.0°F)}	\multicolumn{4}{c}{40 psia (267.3°F)}
Sat.	20.09	1082.0	1156.4	1.7320	10.50	1092.3	1170.0	1.6767
250	20.79	1090.3	1167.2	1.7475				
300	22.36	1108.7	1191.5	1.7805	11.04	1105.1	1186.8	1.6993
350	23.90	1126.9	1215.4	1.8110	11.84	1124.2	1211.8	1.7312
400	25.43	1145.1	1239.2	1.8395	12.62	1143.0	1236.4	1.7606
500	28.46	1181.5	1286.8	1.8919	14.16	1180.1	1284.9	1.8140
600	31.47	1218.4	1334.8	1.9395	15.69	1217.3	1333.4	1.8621
700	34.47	1255.9	1383.5	1.9834	17.20	1255.1	1382.4	1.9063
800	37.46	1294.3	1432.9	2.0243	18.70	1293.7	1432.1	1.9474
900	40.45	1333.5	1483.2	2.0627	20.20	1333.0	1482.5	1.9859
1000	43.44	1373.5	1534.3	2.0989	21.70	1373.1	1533.8	2.0223
1100	46.42	1414.5	1586.3	2.1334	23.20	1414.2	1585.9	2.0568
	\multicolumn{4}{c}{60 psia (292.7°F)}	\multicolumn{4}{c}{80 psia (312.1°F)}
Sat.	7.18	1098.3	1178.0	1.6444	5.47	1102.6	1183.6	1.6214
300	7.26	1101.3	1181.9	1.6496				
350	7.82	1121.4	1208.2	1.6830	5.80	1118.5	1204.3	1.6476
400	8.35	1140.8	1233.5	1.7134	6.22	1138.5	1230.6	1.6790
500	9.40	1178.6	1283.0	1.7678	7.02	1177.2	1281.1	1.7346
600	10.43	1216.3	1332.1	1.8165	7.79	1215.3	1330.7	1.7838
700	11.44	1254.4	1381.4	1.8609	8.56	1253.6	1380.3	1.8285
800	12.45	1293.0	1431.2	1.9022	9.32	1292.4	1430.4	1.8700
900	13.45	1332.5	1481.8	1.9408	10.08	1332.0	1481.2	1.9087
1000	14.45	1372.7	1533.2	1.9773	10.83	1372.3	1532.6	1.9453
1100	15.45	1413.8	1585.4	2.0119	11.58	1413.5	1584.9	1.9799
1200	16.45	1455.8	1638.5	2.0448	12.33	1455.5	1638.1	2.0130

Table B-2C (Continued)

Temp., °F	v	u	h	s	v	u	h	s
	\multicolumn{4}{c}{100 psia (327.9°F)}	\multicolumn{4}{c}{120 psia (341.3°F)}
Sat.	4.434	1105.8	1187.8	1.6034	3.730	1108.3	1191.1	1.5886
350	4.592	1115.4	1200.4	1.6191	3.783	1112.2	1196.2	1.5950
400	4.934	1136.2	1227.5	1.6517	4.079	1133.8	1224.4	1.6288
450	5.265	1156.2	1253.6	1.6812	4.360	1154.3	1251.2	1.6590
500	5.587	1175.7	1279.1	1.7085	4.633	1174.2	1277.1	1.6868
600	6.216	1214.2	1329.3	1.7582	5.164	1213.2	1327.8	1.7371
700	6.834	1252.8	1379.2	1.8033	5.682	1252.0	1378.2	1.7825
800	7.445	1291.8	1429.6	1.8449	6.195	1291.2	1428.7	1.8243
900	8.053	1331.5	1480.5	1.8838	6.703	1330.9	1479.8	1.8633
1000	8.657	1371.9	1532.1	1.9204	7.208	1371.5	1531.5	1.9000
1100	9.260	1413.1	1584.5	1.9551	7.711	1412.8	1584.0	1.9348
1200	9.861	1455.2	1637.7	1.9882	8.213	1454.9	1637.3	1.9679
	\multicolumn{4}{c}{140 psia (353.1°F)}	\multicolumn{4}{c}{160 psia (363.6°F)}
Sat.	3.221	1110.3	1193.8	1.5761	2.836	1112.0	1196.0	1.5651
400	3.466	1131.4	1221.2	1.6088	3.007	1128.8	1217.8	1.5911
450	3.713	1152.4	1248.6	1.6399	3.228	1150.5	1246.1	1.6230
500	3.952	1172.7	1275.1	1.6682	3.440	1171.2	1273.0	1.6518
550	4.184	1192.6	1300.9	1.6944	3.646	1191.3	1299.2	1.6784
600	4.412	1212.1	1326.4	1.7191	3.848	1211.1	1325.0	1.7034
700	4.860	1251.2	1377.1	1.7648	4.243	1250.4	1376.0	1.7494
800	5.301	1290.5	1427.9	1.8068	4.631	1289.9	1427.0	1.7916
900	5.739	1330.4	1479.1	1.8459	5.015	1329.9	1478.4	1.8308
1000	6.173	1371.0	1531.0	1.8827	5.397	1370.6	1530.4	1.8677
1100	6.605	1412.4	1583.6	1.9176	5.776	1412.1	1583.1	1.9026
1200	7.036	1454.6	1636.9	1.9507	6.154	1454.3	1636.5	1.9358
	\multicolumn{4}{c}{180 psia (373.1°F)}	\multicolumn{4}{c}{200 psia (381.9°F)}
Sat.	2.533	1113.4	1197.8	1.5553	2.289	1114.6	1199.3	1.5464
400	2.648	1126.2	1214.4	1.5749	2.361	1123.5	1210.8	1.5600
450	2.850	1148.5	1243.4	1.6078	2.548	1146.4	1240.7	1.5938
500	3.042	1169.6	1270.9	1.6372	2.724	1168.0	1268.8	1.6239
550	3.228	1190.0	1297.5	1.6642	2.893	1188.7	1295.7	1.6512
600	3.409	1210.0	1323.5	1.6893	3.058	1208.9	1322.1	1.6767
700	3.763	1249.6	1374.9	1.7357	3.379	1248.8	1373.8	1.7234
800	4.110	1289.3	1426.2	1.7781	3.693	1288.6	1425.3	1.7660
900	4.453	1329.4	1477.7	1.8175	4.003	1328.9	1477.1	1.8055
1000	4.793	1370.2	1529.8	1.8545	4.310	1369.8	1529.3	1.8425
1100	5.131	1411.7	1582.6	1.8894	4.615	1411.4	1582.2	1.8776
1200	5.467	1454.0	1636.1	1.9227	4.918	1453.7	1635.7	1.9109

Table B-2C (Continued)

Temp., °F	v	u	h	s	v	u	h	s
	250 psia (401.0°F)				300 psia (417.4°F)			
Sat.	1.845	1116.7	1202.1	1.5274	1.544	1118.2	1203.9	1.5115
450	2.002	1141.1	1233.7	1.5632	1.636	1135.4	1226.2	1.5365
500	2.150	1163.8	1263.3	1.5948	1.766	1159.5	1257.5	1.5701
550	2.290	1185.3	1291.3	1.6233	1.888	1181.9	1286.7	1.5997
600	2.426	1206.1	1318.3	1.6494	2.004	1203.2	1314.5	1.6266
700	2.688	1246.7	1371.1	1.6970	2.227	1244.6	1368.3	1.6751
800	2.943	1287.0	1423.2	1.7401	2.442	1285.4	1421.0	1.7187
900	3.193	1327.6	1475.3	1.7799	2.653	1326.3	1473.6	1.7589
1000	3.440	1368.7	1527.9	1.8172	2.860	1367.7	1526.5	1.7964
1100	3.685	1410.5	1581.0	1.8524	3.066	1409.6	1579.8	1.8317
1200	3.929	1453.0	1634.8	1.8858	3.270	1452.2	1633.8	1.8653
1300	4.172	1496.3	1689.3	1.9177	3.473	1495.6	1688.4	1.8973
	350 psia (431.8°F)				400 psia (444.7°F)			
Sat.	1.327	1119.0	1204.9	1.4978	1.162	1119.5	1205.5	1.4856
450	1.373	1129.2	1218.2	1.5125	1.175	1122.6	1209.6	1.4901
500	1.491	1154.9	1251.5	1.5482	1.284	1150.1	1245.2	1.5282
550	1.600	1178.3	1281.9	1.5790	1.383	1174.6	1277.0	1.5605
600	1.703	1200.3	1310.6	1.6068	1.476	1197.3	1306.6	1.5892
700	1.898	1242.5	1365.4	1.6562	1.650	1240.4	1362.5	1.6397
800	2.085	1283.8	1418.8	1.7004	1.816	1282.1	1416.6	1.6844
900	2.267	1325.0	1471.8	1.7409	1.978	1323.7	1470.1	1.7252
1000	2.446	1366.6	1525.0	1.7787	2.136	1365.5	1523.6	1.7632
1100	2.624	1408.7	1578.6	1.8142	2.292	1407.8	1577.4	1.7989
1200	2.799	1451.5	1632.8	1.8478	2.446	1450.7	1631.8	1.8327
1300	2.974	1495.0	1687.6	1.8799	2.599	1494.3	1686.8	1.8648
	450 psia (456.4°F)				500 psia (467.1°F)			
Sat.	1.033	1119.6	1205.6	1.4746	0.928	1119.4	1205.3	1.4645
500	1.123	1145.1	1238.5	1.5097	0.992	1139.7	1231.5	1.4923
550	1.215	1170.7	1271.9	1.5436	1.079	1166.7	1266.6	1.5279
600	1.300	1194.3	1302.5	1.5732	1.158	1191.1	1298.3	1.5585
700	1.458	1238.2	1359.6	1.6248	1.304	1236.0	1356.7	1.6112
800	1.608	1280.5	1414.4	1.6701	1.441	1278.8	1412.1	1.6571
900	1.752	1322.4	1468.3	1.7113	1.572	1321.0	1466.5	1.6987
1000	1.894	1364.4	1522.2	1.7495	1.701	1363.3	1520.7	1.7471
1100	2.034	1406.9	1576.3	1.7853	1.827	1406.0	1575.1	1.7731
1200	2.172	1450.0	1630.8	1.8192	1.952	1449.2	1629.8	1.8072
1300	2.308	1493.7	1685.9	1.8515	2.075	1493.1	1685.1	1.8395
1400	2.444	1538.1	1741.7	1.8823	2.198	1537.6	1741.0	1.8704

Table B-2C (Continued)

Temp., °F	v	u	h	s	v	u	h	s
	\multicolumn{4}{c}{600 psia (486.3°F)}	\multicolumn{4}{c}{700 psia (503.2°F)}
Sat.	0.770	1118.6	1204.1	1.4464	0.656	1117.0	1202.0	1.4305
500	0.795	1128.0	1216.2	1.4592				
550	0.875	1158.2	1255.4	1.4990	0.728	1149.0	1243.2	1.4723
600	0.946	1184.5	1289.5	1.5320	0.793	1177.5	1280.2	1.5081
700	1.073	1231.5	1350.6	1.5872	0.907	1226.9	1344.4	1.5661
800	1.190	1275.4	1407.6	1.6343	1.011	1272.0	1402.9	1.6145
900	1.302	1318.4	1462.9	1.6766	1.109	1315.6	1459.3	1.6576
1000	1.411	1361.2	1517.8	1.7155	1.204	1358.9	1514.9	1.6970
1100	1.517	1404.2	1572.7	1.7519	1.296	1402.4	1570.2	1.7337
1200	1.622	1447.7	1627.8	1.7861	1.387	1446.2	1625.8	1.7682
1300	1.726	1491.7	1683.4	1.8186	1.476	1490.4	1681.7	1.8009
1400	1.829	1536.5	1739.5	1.8497	1.565	1535.3	1738.1	1.8321
	\multicolumn{4}{c}{800 psia (518.4°F)}	\multicolumn{4}{c}{900 psia (532.1°F)}
Sat.	0.569	1115.0	1199.3	1.4160	0.501	1112.6	1196.0	1.4027
550	0.615	1138.8	1229.9	1.4469	0.527	1127.5	1215.2	1.4219
600	0.677	1170.1	1270.4	1.4861	0.587	1162.2	1260.0	1.4652
650	0.732	1197.2	1305.6	1.5186	0.639	1191.1	1297.5	1.4999
700	0.783	1222.1	1338.0	1.5471	0.686	1217.1	1331.4	1.5297
800	0.876	1268.5	1398.2	1.5969	0.772	1264.9	1393.4	1.5810
900	0.964	1312.9	1455.6	1.6408	0.851	1310.1	1451.9	1.6257
1000	1.048	1356.7	1511.9	1.6807	0.927	1354.5	1508.9	1.6662
1100	1.130	1400.5	1567.8	1.7178	1.001	1398.7	1565.4	1.7036
1200	1.210	1444.6	1623.8	1.7526	1.073	1443.0	1621.7	1.7386
1300	1.289	1489.1	1680.0	1.7854	1.144	1487.8	1687.3	1.7717
1400	1.367	1534.2	1736.6	1.8167	1.214	1533.0	1735.1	1.8031
	\multicolumn{4}{c}{1000 psia (544.8°F)}	\multicolumn{4}{c}{1200 psia (567.4°F)}
Sat.	0.446	1109.0	1192.4	1.3903	0.362	1103.5	1183.9	1.3673
600	0.514	1153.7	1248.8	1.4450	0.402	1134.4	1223.6	1.4054
650	0.564	1184.7	1289.1	1.4822	0.450	1170.9	1270.8	1.4490
700	0.608	1212.0	1324.6	1.5135	0.491	1201.3	1310.2	1.4837
800	0.688	1261.2	1388.5	1.5664	0.562	1253.7	1378.4	1.5402
900	0.761	1307.3	1448.1	1.6120	0.626	1301.5	1440.4	1.5876
1000	0.831	1352.2	1505.9	1.6530	0.685	1347.5	1499.7	1.6297
1100	0.898	1396.8	1562.9	1.6908	0.743	1393.0	1557.9	1.6682
1200	0.963	1441.5	1619.7	1.7261	0.798	1438.3	1615.5	1.7040
1300	1.027	1486.5	1676.5	1.7593	0.853	1483.8	1673.1	1.7377
1400	1.091	1531.9	1733.7	1.7909	0.906	1529.6	1730.7	1.7696
1600	1.215	1624.4	1849.3	1.8499	1.011	1622.6	1847.1	1.8290

Table B-2D

Properties of Water: Compressed Liquid Table v, ft^3/lb; h, Btu/lb; s, Btu/lb·°R

Temp., °F	500 psia (T_{sat} = 467.1°F)				1000 psia (T_{sat} = 544.8°F)			
	v	u	h	s	v	u	h	s
32	0.015994	0.00	1.49	0.00000	0.015967	0.03	2.99	0.00005
50	0.015998	18.02	19.50	0.03599	0.015972	17.99	20.94	0.03592
100	0.016106	67.87	69.36	0.12932	0.016082	67.70	70.68	0.12901
150	0.016318	117.66	119.17	0.21457	0.016293	117.38	120.40	0.21410
200	0.016608	167.65	169.19	0.29341	0.016580	167.26	170.32	0.29281
300	0.017416	268.92	270.53	0.43641	0.017379	268.24	271.46	0.43552
400	0.018608	373.68	375.40	0.56604	0.018550	372.55	375.98	0.56472
Sat.	0.019748	447.70	449.53	0.64904	0.021591	538.39	542.38	0.74320
	1500 psia (T_{sat} = 596.4°F)				2000 psia (T_{sat} = 636.0°F)			
32	0.015939	0.05	4.47	0.00007	0.015912	0.06	5.95	0.00008
50	0.015946	17.95	22.38	0.03584	0.015920	17.91	23.81	0.03575
100	0.016058	67.53	71.99	0.12870	0.016034	67.37	73.30	0.12839
150	0.016268	117.10	121.62	0.21364	0.016244	116.83	122.84	0.21318
200	0.016554	166.87	171.46	0.29221	0.016527	166.49	172.60	0.29162
300	0.017343	267.58	272.39	0.43463	0.017308	266.93	273.33	0.43376
400	0.018493	371.45	376.59	0.56343	0.018439	370.38	377.21	0.56216
500	0.02024	481.8	487.4	0.6853	0.02014	479.8	487.3	0.6832
Sat.	0.02346	605.0	611.5	0.8082	0.02565	662.4	671.9	0.8623
	3000 psia (T_{sat} = 695.5°F)				4000 psia			
32	0.015859	0.09	8.90	0.00009	0.015807	0.10	11.80	0.00005
50	0.015870	17.84	26.65	0.03555	0.015821	17.76	29.47	0.03534
100	0.015987	67.04	75.91	0.12777	0.015942	66.72	78.52	0.12714
150	0.016196	116.30	125.29	0.21226	0.016150	115.77	127.73	0.21136
200	0.016476	165.74	174.89	0.29046	0.016425	165.02	177.18	0.28931
300	0.017240	265.66	275.23	0.43205	0.017174	264.43	277.15	0.43038
400	0.018334	368.32	378.50	0.55970	0.018235	366.35	379.85	0.55734
500	0.019944	476.2	487.3	0.6794	0.019766	472.9	487.5	0.6758
Sat.	0.034310	783.5	802.5	0.9732				

Table B-3
Properties of Refrigerant 12 (CCl_2F_2)[a]

Table B-3A
Properties of Saturated Refrigerant 12: Temperature Table
v, ft^3/lb; u, Btu/lb; h, Btu/lb; s, Btu/lb°R

		Specific volume		Internal energy		Enthalpy			Entropy	
Temp. °F T	Press. psi P	Sat. liquid v_f	Sat. vapor v_g	Sat. liquid u_f	Sat. vapor u_g	Sat. liquid h_f	Evap. h_{fg}	Sat. vapor h_g	Sat. liquid s_f	Sat. vapor s_g
−40	9.308	0.01056	3.8750	−0.02	66.24	0	72.91	72.91	0	0.1737
−30	11.999	0.01067	3.0585	1.93	67.22	2.11	71.90	74.02	0.0050	0.1723
−20	15.267	0.01079	2.4429	4.21	68.21	4.24	70.87	75.11	0.0098	0.1710
−10	19.189	0.01091	1.9727	6.33	69.19	6.37	69.82	76.20	0.0146	0.1699
0	23.849	0.01103	1.6089	8.47	70.17	8.52	68.75	77.27	0.0193	0.1689
10	29.335	0.01116	1.3241	10.62	71.15	10.68	67.65	78.34	0.0240	0.1680
20	35.736	0.01130	1.0988	12.79	72.12	12.86	66.52	79.39	0.0285	0.1672
30	43.148	0.01144	0.9188	14.97	73.08	15.06	65.36	80.42	0.0330	0.1665
40	51.667	0.01159	0.7736	17.16	74.04	17.27	64.16	81.44	0.0375	0.1659
50	61.394	0.01175	0.6554	19.38	74.99	19.51	62.93	82.43	0.0418	0.1653
60	72.433	0.01191	0.5584	21.61	75.92	21.77	61.64	83.41	0.0462	0.1648
70	84.888	0.01209	0.4782	23.86	76.85	24.05	60.31	84.36	0.0505	0.1643
80	98.870	0.01228	0.4114	26.14	77.76	26.37	58.92	85.28	0.0548	0.1639
90	114.49	0.01248	0.3553	28.45	78.65	28.71	57.46	86.17	0.0590	0.1635
100	131.86	0.01269	0.3079	30.79	79.51	31.10	55.93	87.03	0.0632	0.1632
110	151.11	0.01292	0.2677	33.16	80.36	33.53	54.31	87.84	0.0675	0.1628
120	172.35	0.01317	0.2333	35.59	81.17	36.01	52.60	88.61	0.0717	0.1624
140	221.32	0.01375	0.1780	40.60	82.68	41.16	48.81	89.97	0.0802	0.1616
160	279.82	0.01445	0.1360	45.88	83.96	46.63	44.37	91.01	0.0889	0.1605
180	349.00	0.01536	0.1033	51.57	84.89	52.56	39.00	91.56	0.0980	0.1590
200	430.09	0.01666	0.0767	57.87	85.17	59.20	32.08	91.28	0.1079	0.1565
233.6	596.9	0.0287	0.0287	75.69	75.69	78.86	0	78.86	0.1359	0.1359

[a] Data from Freon Products Division, E.I. du Pont de Nemours & Company, 1956.

Table B-3B

Properties of Saturated Refrigerant 12: Pressure Table
v, ft³/lb; u, Btu/lb; h, Btu/lb; s, Btu/lb°R

Press. psi P	Temp. °F T	Specific volume		Internal energy		Enthalpy			Entropy	
		Sat. liquid v_f	Sat. vapor v_g	Sat. liquid u_f	Sat. vapor u_g	Sat. liquid h_f	Evap. h_{fg}	Sat. vapor h_g	Sat. liquid s_f	Sat. vapor s_g
5	−62.35	0.0103	6.9069	−4.69	64.04	−4.68	75.11	70.43	−0.0114	0.1776
10	−37.23	0.0106	3.6246	0.56	66.51	0.58	72.64	73.22	0.0014	0.1733
15	−20.75	0.0108	2.4835	4.05	68.13	4.08	70.95	75.03	0.0095	0.1711
20	−8.13	0.0109	1.8977	6.73	69.37	6.77	69.63	76.40	0.0155	0.1697
30	11.11	0.0112	1.2964	10.86	71.25	10.93	67.53	78.45	0.0245	0.1679
40	25.93	0.0114	0.9874	14.08	72.69	14.16	65.84	80.00	0.0312	0.1668
50	38.15	0.0116	0.7982	16.75	73.86	16.86	64.39	81.25	0.0366	0.1660
60	48.64	0.0117	0.6701	19.07	74.86	19.20	63.10	82.30	0.0413	0.1654
70	57.90	0.0119	0.5772	21.13	75.73	21.29	61.92	83.21	0.0453	0.1649
80	66.21	0.0120	0.5068	23.00	76.50	23.18	60.82	84.00	0.0489	0.1645
90	73.79	0.0122	0.4514	24.72	77.20	24.92	59.79	84.71	0.0521	0.1642
100	80.76	0.0123	0.4067	26.31	77.82	26.54	58.81	85.35	0.0551	0.1639
120	93.29	0.0126	0.3389	29.21	78.93	29.49	56.97	86.46	0.0604	0.1634
140	104.35	0.0128	0.2896	31.82	79.89	32.15	55.24	87.39	0.0651	0.1630
160	114.30	0.0130	0.2522	34.21	80.71	34.59	53.59	88.18	0.0693	0.1626
180	123.38	0.0133	0.2228	36.42	81.44	36.86	52.00	88.86	0.0731	0.1623
200	131.74	0.0135	0.1989	38.50	82.08	39.00	50.44	89.44	0.0767	0.1620
220	139.51	0.0137	0.1792	40.48	82.08	41.03	48.90	89.94	0.0816	0.1616
240	146.77	0.0140	0.1625	42.35	83.14	42.97	47.39	90.36	0.0831	0.1613
260	153.60	0.0142	0.1483	44.16	83.58	44.84	45.88	90.72	0.0861	0.1609
280	160.06	0.0145	0.1359	45.90	83.97	46.65	44.36	91.01	0.0890	0.1605
300	166.18	0.0147	0.1251	47.59	84.30	48.41	42.83	91.24	0.0917	0.1601

Table B-3C

Properties of Superheated Refrigerant 12
v, ft^3/lb; u, Btu/lb; h, Btu/lb; s, Btu/lb°R

Temp. °F	v	u	h	s	v	u	h	s
	\multicolumn{4}{c}{10 psia (−37.23°F)}	\multicolumn{4}{c}{15 psia (−20.75°F)}
Sat.	3.6246	66.512	73.219	0.1733	2.4835	68.134	75.028	0.1711
0	3.9809	70.879	78.246	0.1847	2.6201	70.629	77.902	0.1775
20	4.1691	73.299	81.014	0.1906	2.7494	73.080	80.712	0.1835
40	4.3556	75.768	83.828	0.1964	2.8770	75.575	83.561	0.1893
60	4.5408	78.286	86.689	0.2020	3.0031	78.115	86.451	0.1950
80	4.7248	80.853	89.596	0.2075	3.1281	80.700	89.383	0.2005
100	4.9079	83.466	92.548	0.2128	3.2521	83.330	92.357	0.2059
120	5.0903	86.126	95.546	0.2181	3.3754	86.004	95.373	0.2112
140	5.2720	88.830	98.586	0.2233	3.4981	88.719	98.429	0.2164
160	5.4533	91.578	101.669	0.2283	3.6202	91.476	101.525	0.2215
180	5.6341	94.367	104.793	0.2333	3.7419	94.274	104.661	0.2265
200	5.8145	97.197	107.957	0.2381	3.8632	97.112	107.835	0.2314
	\multicolumn{4}{c}{20 psia (−8.13°F)}	\multicolumn{4}{c}{30 psia (11.11°F)}
Sat.	1.8977	69.374	76.397	0.1697	1.2964	71.255	78.452	0.1679
20	2.0391	72.856	80.403	0.1783	1.3278	72.394	79.765	0.1707
40	2.1373	75.379	83.289	0.1842	1.3969	74.975	82.730	0.1767
60	2.2340	77.942	86.210	0.1899	1.4644	77.586	85.716	0.1826
80	2.3295	80.546	89.168	0.1955	1.5306	80.232	88.729	0.1883
100	2.4241	83.192	92.164	0.2010	1.5957	82.911	91.770	0.1938
120	2.5179	85.879	95.198	0.2063	1.6600	85.627	94.843	0.1992
140	2.6110	88.607	98.270	0.2115	1.7237	88.379	97.948	0.2045
160	2.7036	91.374	101.380	0.2166	1.7868	91.166	101.086	0.2096
180	2.7957	94.181	104.528	0.2216	1.8494	93.991	104.258	0.2146
200	2.8874	97.026	107.712	0.2265	1.9116	96.852	107.464	0.2196
220	2.9789	99.907	110.932	0.2313	1.9735	99.746	110.702	0.2244
	\multicolumn{4}{c}{40 psia (25.93°F)}	\multicolumn{4}{c}{50 psia (38.15°F)}
Sat.	0.9874	72.691	80.000	0.1668	0.7982	73.863	81.249	0.1660
40	1.0258	74.555	82.148	0.1711	0.8025	74.115	81.540	0.1666
60	1.0789	77.220	85.206	0.1771	0.8471	76.838	84.676	0.1727
80	1.1306	79.908	88.277	0.1829	0.8903	79.574	87.811	0.1786
100	1.1812	82.624	91.367	0.1885	0.9322	82.328	90.953	0.1843
120	1.2309	85.369	94.480	0.1940	0.9731	85.106	94.110	0.1899
140	1.2798	88.147	97.620	0.1993	1.0133	87.910	97.286	0.1953
160	1.3282	90.957	100.788	0.2045	1.0529	90.743	100.485	0.2005
180	1.3761	93.800	103.985	0.2096	1.0920	93.604	103.708	0.2056
200	1.4236	96.674	107.212	0.2146	1.1307	96.496	106.958	0.2106
220	1.4707	99.583	110.469	0.2194	1.1690	99.419	110.235	0.2155
240	1.5176	102.524	113.757	0.2242	1.2070	102.371	113.539	0.2203

Table B3-C (Continued)

Temp., °F	v	u	h	s	v	u	h	s
	60 psia (48.64°F)				70 psia (57.90°F)			
Sat.	0.6701	74.859	82.299	0.1654	0.5772	75.729	83.206	0.1649
60	0.6921	76.442	84.126	0.1689	0.5809	76.027	83.552	0.1656
80	0.7296	79.229	87.330	0.1750	0.6146	78.871	86.832	0.1718
100	0.7659	82.024	90.528	0.1808	0.6469	81.712	90.091	0.1777
120	0.8011	84.836	93.731	0.1864	0.6780	84.560	93.343	0.1834
140	0.8355	87.668	96.945	0.1919	0.7084	87.421	96.597	0.1889
160	0.8693	90.524	100.176	0.1972	0.7380	90.302	99.862	0.1943
180	0.9025	93.406	103.427	0.2023	0.7671	93.205	103.141	0.1995
200	0.9353	96.315	106.700	0.2074	0.7957	96.132	106.439	0.2046
220	0.9678	99.252	109.997	0.2123	0.8240	99.083	109.756	0.2095
240	0.9998	102.217	113.319	0.2171	0.8519	102.061	113.096	0.2144
260	1.0318	105.210	116.666	0.2218	0.8796	105.065	116.459	0.2191
	80 psia (66.21°F)				90 psia (73.79°F)			
Sat.	0.5068	76.500	84.003	0.1645	0.4514	77.194	84.713	0.1642
80	0.5280	78.500	86.316	0.1689	0.4602	78.115	85.779	0.1662
100	0.5573	81.389	89.640	0.1749	0.4875	81.056	89.175	0.1723
120	0.5856	84.276	92.945	0.1807	0.5135	83.984	92.536	0.1782
140	0.6129	87.169	96.242	0.1863	0.5385	86.911	95.879	0.1839
160	0.6394	90.076	99.542	0.1917	0.5627	89.845	99.216	0.1894
180	0.6654	93.000	102.851	0.1970	0.5863	92.793	102.557	0.1947
200	0.6910	95.945	106.174	0.2021	0.6094	95.755	105.905	0.1998
220	0.7161	98.912	109.513	0.2071	0.6321	98.739	109.267	0.2049
240	0.7409	101.904	112.872	0.2119	0.6545	101.743	112.644	0.2098
260	0.7654	104.919	116.251	0.2167	0.6766	104.771	116.040	0.2146
280	0.7898	107.960	119.652	0.2214	0.6985	107.823	119.456	0.2192
	100 psia (80.76°F)				120 psia (93.29°F)			
Sat.	0.4067	77.824	85.351	0.1639	0.3389	78.933	86.459	0.1634
100	0.4314	80.711	88.694	0.1700	0.3466	79.978	87.675	0.1656
120	0.4556	83.685	92.116	0.1760	0.3684	83.056	91.237	0.1718
140	0.4788	86.647	95.507	0.1817	0.3890	86.098	94.736	0.1778
160	0.5012	89.610	98.884	0.1873	0.4087	89.123	98.199	0.1835
180	0.5229	92.580	102.257	0.1926	0.4277	92.144	101.642	0.1889
200	0.5441	95.564	105.633	0.1978	0.4461	95.170	105.076	0.1942
220	0.5649	98.564	109.018	0.2029	0.4640	98.205	108.509	0.1993
240	0.5854	101.582	112.415	0.2078	0.4816	101.253	111.948	0.2043
260	0.6055	104.622	115.828	0.2126	0.4989	104.317	115.396	0.2092
280	0.6255	107.684	119.258	0.2173	0.5159	107.401	118.857	0.2139
300	0.6452	110.768	122.707	0.2219	0.5327	110.504	122.333	0.2186

Table B-3C (Continued)

Temp. °F	v	u	h	s	v	u	h	s
	140 psia (104.35°F)				160 psia (114.30°F)			
Sat.	0.2896	79.886	87.389	0.1630	0.2522	80.713	88.180	0.1626
120	0.3055	82.382	90.297	0.1681	0.2576	81.656	89.283	0.1645
140	0.3245	85.516	93.923	0.1742	0.2756	84.899	93.059	0.1709
160	0.3423	88.615	97.483	0.1801	0.2922	88.080	96.732	0.1770
180	0.3594	91.692	101.003	0.1857	0.3080	91.221	100.340	0.1827
200	0.3758	94.765	104.501	0.1910	0.3230	94.344	103.907	0.1882
220	0.3918	97.837	107.987	0.1963	0.3375	97.457	107.450	0.1935
240	0.4073	100.918	111.470	0.2013	0.3516	100.570	110.980	0.1986
260	0.4226	104.008	114.956	0.2062	0.3653	103.690	114.506	0.2036
280	0.4375	107.115	118.449	0.2110	0.3787	106.820	118.033	0.2084
300	0.4523	110.235	121.953	0.2157	0.3919	109.964	121.567	0.2131
320	0.4668	113.376	125.470	0.2202	0.4049	113.121	125.109	0.2177
	180 psia (123.38°F)				200 psia (131.74°F)			
Sat.	0.2228	81.436	88.857	0.1623	0.1989	82.077	89.439	0.1620
140	0.2371	84.238	92.136	0.1678	0.2058	83.521	91.137	0.1648
160	0.2530	87.513	95.940	0.1741	0.2212	86.913	95.100	0.1713
180	0.2678	90.727	99.647	0.1800	0.2354	90.211	98.921	0.1774
200	0.2818	93.904	103.291	0.1856	0.2486	93.451	102.652	0.1831
220	0.2952	97.063	106.896	0.1910	0.2612	96.659	106.325	0.1886
240	0.3081	100.215	110.478	0.1961	0.2732	99.850	109.962	0.1939
260	0.3207	103.364	114.046	0.2012	0.2849	103.032	113.576	0.1990
280	0.3329	106.521	117.610	0.2061	0.2962	106.214	117.178	0.2039
300	0.3449	109.686	121.174	0.2108	0.3073	109.402	120.775	0.2087
320	0.3567	112.863	124.744	0.2155	0.3182	112.598	124.373	0.2134
340	0.3683	116.053	128.321	0.2200	0.3288	115.805	127.974	0.2179
	300 psia (166.18°F)				400 psia (192.93°F)			
Sat.	0.1251	84.295	91.240	0.1601	0.0856	85.178	91.513	0.1576
180	0.1348	87.071	94.556	0.1654				
200	0.1470	90.816	98.975	0.1722	0.0910	86.982	93.718	0.1609
220	0.1577	94.379	103.136	0.1784	0.1032	91.410	99.046	0.1689
240	0.1676	97.835	107.140	0.1842	0.1130	95.371	103.735	0.1757
260	0.1769	101.225	111.043	0.1897	0.1216	99.102	108.105	0.1818
280	0.1856	104.574	114.879	0.1950	0.1295	102.701	112.286	0.1876
300	0.1940	107.899	118.670	0.2000	0.1368	106.217	116.343	0.1930
320	0.2021	111.208	122.430	0.2049	0.1437	109.680	120.318	0.1981
340	0.2100	114.512	126.171	0.2096	0.1503	113.108	124.235	0.2031
360	0.2177	117.814	129.900	0.2142	0.1567	116.514	128.112	0.2079

Table B-4
Thermodynamic Properties of Nitrogen[a]

Specific volume: v, ft³/lb, enthalpy: h, Btu/lb, entropy: s, Btu/lb°R

Table B-4A
Saturated Nitrogen: Temperature Table

Temperature °R	Absolute Pressure lbf/in.² p	Specific Volume			Enthalpy			Entropy		
		Sat. liquid v_f	v_{fg}	Sat. Vapor v_g	Sat. liquid h_f	h_{fg}	Sat. Vapor h_g	Sat. Liquid s_f	s_{fg}	Sat. Vapor s_g
113.670	1.813	0.01845	23.793	23.812	0.000	92.891	92.891	0.00000	0.81720	0.81720
120.000	3.337	0.01875	13.570	13.589	3.113	91.224	94.337	0.02661	0.76020	0.78681
130.000	7.654	0.01929	6.3208	6.3401	8.062	88.432	96.494	0.06610	0.68025	0.74634
139.255	14.696	0.01984	3.4592	3.4791	12.639	85.668	98.306	0.09992	0.61518	0.71510
140.000	15.425	0.01989	3.3072	3.3271	13.006	85.436	98.443	0.10253	0.61026	0.71279
150.000	28.120	0.02056	1.8865	1.9071	17.945	82.179	100.124	0.13628	0.54786	0.68414
160.000	47.383	0.02132	1.1469	1.1682	22.928	78.548	101.476	0.16795	0.49093	0.65888
170.000	74.991	0.02219	0.7299	0.7521	28.045	74.383	102.427	0.19829	0.43754	0.63584
180.000	112.808	0.02323	0.4789	0.5021	33.411	69.478	102.889	0.22805	0.38599	0.61404
190.000	162.761	0.02449	0.3190	0.3435	39.153	63.582	102.735	0.25789	0.33464	0.59254
200.000	226.853	0.02613	0.2119	0.2380	45.283	56.474	101.757	0.28780	0.28237	0.57017
210.000	307.276	0.02845	0.1354	0.1639	52.061	47.474	99.536	0.31894	0.22607	0.54501
220.000	406.739	0.03249	0.0750	0.1075	60.336	34.536	94.872	0.35494	0.15698	0.51192
226.000	477.104	0.03806	0.0374	0.0755	68.123	20.423	88.546	0.38789	0.09037	0.47826

[a] Abstracted from National Bureau of Standards Technical Note 129A, "The Thermodynamic Properties of Nitrogen from 114 to 540°R between 1.0 and 3000 psia," Supplement A (British Units) by T. R. Strobridge.

Table B-4B

Superheated Nitrogen Table

Tempera-ture °R	v	h	s	v	h	s	v	h	s
	14.7 lbf/in.^{2}			20 lbf/in.^{2}			50 lbf/in.^{2}		
150	3.7782	101.086	0.7343	2.7395	100.715	0.7109			
200	5.1366	113.849	0.8078	3.7538	113.625	0.7852	1.4534	112.315	0.7159
250	6.4680	126.443	0.8640	4.7397	126.293	0.8418	1.8663	125.432	0.7744
300	7.7876	138.958	0.9096	5.7138	138.850	0.8875	2.2662	138.239	0.8212
350	9.1015	151.432	0.9481	6.6820	151.351	0.9261	2.6599	150.896	0.8602
400	10.412	163.882	0.9814	7.6469	163.821	0.9594	3.0502	163.471	0.8938
450	11.721	176.319	1.0107	8.6098	176.271	0.9887	3.4385	175.997	0.9233
500	13.028	188.748	1.0368	9.5714	188.710	1.0149	3.8255	188.492	0.9496
540	14.073	198.690	1.0560	10.340	198.657	1.0341	4.1344	198.474	0.9688
	100 lbf/in.^{2}			200 lbf/in.^{2}			500 lbf/in.^{2}		
200	0.6834	109.931	0.6585	0.2884	103.911	0.5875			
250	0.9078	123.948	0.7212	0.4272	120.763	0.6631	0.1321	108.378	0.5608
300	1.1169	137.205	0.7696	0.5420	135.076	0.7153	0.1966	128.168	0.6335
350	1.3192	150.133	0.8094	0.6490	148.589	0.7570	0.2473	143.838	0.6819
400	1.5181	162.888	0.8435	0.7522	161.718	0.7921	0.2932	158.205	0.7202
450	1.7149	175.540	0.8733	0.8532	174.630	0.8225	0.3368	171.933	0.7526
500	1.9103	188.129	0.8998	0.9529	187.408	0.8494	0.3790	185.292	0.7807
540	2.0660	198.170	0.9192	1.0319	197.567	0.8690	0.4120	195.807	0.8010
	1000 lbf/in.^{2}			2000 lbf/in.^{2}			3000 lbf/in.^{2}		
250	0.0384	78.126	0.4145	0.0286	70.290	0.3596	0.0261	69.719	0.3371
300	0.0828	115.224	0.5514	0.0398	97.820	0.4599	0.0321	93.216	0.4228
350	0.1150	135.789	0.6150	0.0552	122.614	0.5366	0.0403	116.066	0.4933
400	0.1417	152.487	0.6597	0.0699	142.869	0.5908	0.0493	136.883	0.5490
450	0.1659	167.637	0.6954	0.0833	160.406	0.6321	0.0582	155.522	0.5930
500	0.1887	181.969	0.7256	0.0958	176.411	0.6659	0.0667	172.551	0.6289
540	0.2063	193.069	0.7470	0.1053	188.526	0.6892	0.0732	185.361	0.6535

Table B-5

Ideal Gas Specific Heat $\bar{c}_p(T)^a$

$$\bar{c}_p = A + BT + CT^2 + DT^3 \quad (T, K)$$
$$= cal/(gmole\ K)$$

Gas or vapor†	A	$B\times(10)^3$	$C\times(10)^6$	$D\times(10)^9$	Range, K	Maximum % error
Air	6.713	0.4697	1.147	−0.4696	273–1800	0.72
	6.557	1.477	−0.2148	0	273–3800	1.64
CO	6.726	0.4001	1.283	−0.5307	273–1800	0.89
	6.480	1.566	−0.2387	0	273–3800	1.86
CO_2	5.316	14.285	−8.362	1.784	273–1800	0.67
$c_p =$	18.036 −	0.00004474T −	158.08$(T)^{-\frac{1}{2}}$		273–3800	2.65
H_2	6.952	−0.4576	0.9563	−0.2079	273–1800	1.01
	6.424	1.039	−0.07804	0	273–3800	2.14
H_2O	7.700	0.4594	2.521	−0.8587	273–1800	0.53
	6.970	3.464	−0.4833	0	273–3800	2.03
O_2	6.085	3.631	−1.709	0.3133	273–1800	1.19
	6.732	1.505	−0.1791	0	273–3800	3.24
N_2	6.903	−0.3753	1.930	−6.861	273–1800	0.59
	6.529	1.488	−0.2271	0	273–3800	2.05
NH_3	6.5846	6.1251	2.3663	−1.5981	273–1500	0.91
CH_4	4.750	12.00	3.030	−2.630	273–1500	1.33
C_3H_8	−0.966	72.79	−37.55	7.580	273–1500	0.40
C_4H_{10}	0.945	88.73	−43.80	8.360	273–1500	0.54
C_6H_6	−8.650	115.78	−75.40	18.54	273–1500	0.34
C_2H_2	5.21	22.008	−15.59	4.349	273–1500	1.46
CH_3OH	4.55	21.86	−2.91	−1.92	273–1000	0.18

[a]From E.F. Obert, *Concepts of Thermodynamics*, McGraw-Hill, New York, 1960; based on data presented by K.A. Kobe, "Thermochemistry for the Petrochemical Industry," *Petrol. Refiner*, January 1949–November 1954.

Table B-6
Ideal Gas Properties of Selected Gases[a]

Table B-6A
Ideal Gas Properties of Air h and u, Btu/lb, s' Btu/lb°R

T, °R	h	u	s'	T, °R	h	u	s'
300	71.61	51.04	0.46007	1000	240.98	172.43	0.75042
320	76.40	54.46	0.47550	1100	265.99	190.58	0.77426
340	81.18	57.87	0.49002	1200	291.30	209.05	0.79628
360	85.97	61.29	0.50369	1300	316.94	227.83	0.81680
380	90.75	64.70	0.51663	1400	342.90	246.93	0.83604
400	95.53	68.11	0.52890	1500	369.17	266.34	0.85416
420	100.32	71.52	0.54058	1600	395.74	286.06	0.87130
440	105.11	74.93	0.55172	1700	422.59	306.06	0.88758
460	109.90	78.36	0.56235	1800	449.71	326.32	0.90308
480	114.69	81.77	0.57255	1900	477.09	346.85	0.91788
500	119.48	85.20	0.58233	2000	504.71	367.61	0.93205
520	124.27	88.62	0.59173	2100	532.55	388.60	0.94564
537	128.34	91.53	0.59945	2200	560.59	409.78	0.95868
540	129.06	92.04	0.60078	2300	588.82	431.16	0.97123
560	133.86	95.47	0.60950	2400	617.22	452.70	0.98331
580	138.66	98.90	0.61793	2500	645.78	474.40	0.99497
600	143.47	102.34	0.62607	2600	674.49	496.26	1.00623
620	148.28	105.78	0.63395	2700	703.35	518.26	1.01712
640	153.09	109.21	0.64159	2800	732.33	540.40	1.02767
660	157.92	112.67	0.64902	2900	761.45	562.66	1.03788
680	162.73	116.12	0.65621	3000	790.68	585.04	1.04779
700	167.56	119.58	0.66321	3100	820.03	607.53	1.05741
720	172.39	123.04	0.67002	3200	849.48	630.12	1.06676
740	177.23	126.51	0.67665	3300	879.02	652.81	1.07585
760	182.08	129.99	0.68312	3400	908.66	675.60	1.08470
780	186.94	133.47	0.68942	3500	938.40	698.48	1.09332
800	191.81	136.97	0.69558	3600	968.21	721.44	1.10172
820	196.69	140.47	0.70160	3700	998.11	744.48	1.10991
840	201.56	143.98	0.70747	3800	1028.09	767.60	1.11791
860	206.46	147.50	0.71323	3900	1058.14	790.80	1.12571
880	211.35	151.02	0.71886	4000	1088.26	814.06	1.13334
900	216.26	154.57	0.72438	4100	1118.5	837.4	1.14079
920	221.18	158.12	0.72979	4200	1148.7	860.8	1.14809
940	226.11	161.68	0.73509	4300	1179.0	884.3	1.15522
960	231.06	165.26	0.74030	4400	1209.4	907.8	1.16221
980	236.02	168.83	0.74540	4500	1239.9	931.4	1.16905
				4600	1270.4	955.0	1.17575
				4700	1300.9	978.7	1.18232
				4800	1331.5	1002.5	1.18876
				4900	1362.2	1026.3	1.19508
				5000	1392.9	1050.1	1.20129
				5100	1423.6	1074.0	1.20738
				5200	1454.4	1098.0	1.21336
				5300	1485.3	1122.0	1.21923

[a] Data extracted from J. H. Keenan and J. Kaye, "Gas Tables," John Wiley and Sons, New York, 1945. Reprinted by permission of John Wiley & Sons, Inc.

Table B-6B

Ideal Gas Enthalpy, Internal Energy, and Absolute Entropy of Diatomic Nitrogen, N_2
$\bar{h}$ and $\bar{u}$, Btu/lbmol; $\bar{s}'$, Btu/lbmol°R

T, °R	$\bar{h}$	$\bar{u}$	$\bar{s}'$	T, °R	$\bar{h}$	$\bar{u}$	$\bar{s}'$
300	2082.0	1486.2	41.695	1400	9897	7117	52.551
320	2221.0	1585.5	42.143	1500	10649	7670	53.071
340	2360.0	1684.8	42.564	1600	11410	8232	53.561
360	2498.9	1784.0	42.962	1700	12179	8803	54.028
380	2638.0	1883.4	43.337	1800	12956	9382	54.472
400	2777.0	1982.6	43.694	1900	13742	9968	54.896
420	2916.1	2082.0	44.034	2000	14534	10563	55.303
440	3055.1	2181.3	44.357	2100	15334	11164	55.694
460	3194.1	2280.6	44.665	2200	16140	11771	56.068
480	3333.1	2379.9	44.962	2300	16951	12384	56.429
500	3472.2	2479.3	45.246	2400	17768	13002	56.777
520	3611.3	2578.6	45.519	2500	18590	13625	57.112
537	3729.5	2663.1	45.743	2600	19416	14253	57.436
540	3750.3	2678.0	45.781	2700	20246	14885	57.750
560	3889.5	2777.4	46.034	2800	21081	15521	58.053
580	4028.7	2876.9	46.278	2900	21920	16161	58.348
600	4167.9	2976.4	46.514	3000	22762	16804	58.632
620	4307.1	3075.9	46.742	3100	23607	17451	58.910
640	4446.4	3175.5	46.964	3200	24455	18100	59.179
660	4585.8	3275.2	47.178	3300	25306	18753	59.442
680	4725.3	3374.9	47.386	3400	26160	19408	59.697
700	4864.9	3474.8	47.588	3500	27016	20065	59.944
720	5004.5	3574.7	47.785	3600	27874	20725	60.186
740	5144.3	3674.7	47.977	3700	28735	21387	60.422
760	5284.1	3774.9	48.164	3800	29598	22052	60.652
780	5424.2	3875.2	48.345	3900	30463	22718	60.877
800	5564.4	3975.7	48.522	4000	31329	23386	61.097
820	5704.7	4076.3	48.696	4100	32198	24056	61.310
840	5845.3	4177.1	48.865	4200	33068	24728	61.520
860	5985.9	4278.1	49.031	4300	33940	25401	61.726
880	6126.9	4379.4	49.193	4400	34813	26075	61.927
900	6268.1	4480.8	49.352	4500	35688	26751	62.123
920	6409.6	4582.6	49.507	4600	36564	27429	62.316
940	6551.2	4684.5	49.659	4700	37441	28108	62.504
960	6693.1	4786.7	49.808	4800	38320	28787	62.689
980	6835.4	4889.3	49.955	4900	39199	29468	62.870
1000	6978	4992	50.099	5000	40080	30151	63.049
1100	7695	5511	50.783	5100	40962	30834	63.223
1200	8420	6037	51.413	5200	41844	31518	63.395
1300	9154	6572	52.001	5300	42728	32203	63.563

Table B-6C

Ideal Gas Enthalpy, Internal Energy, and Absolute Entropy of Diatomic Oxygen, O_2
$\bar{h}$ and $\bar{u}$, Btu/lbmol; $\bar{s}'$, Btu/lbmol°R

T, °R	$\bar{h}$	$\bar{u}$	$\bar{s}'$	T, °R	$\bar{h}$	$\bar{u}$	$\bar{s}'$
300	2073.5	1477.8	44.927	1400	10210	7430	56.099
320	2212.6	1577.1	45.375	1500	11017	8038	56.656
340	2351.7	1676.5	45.797	1600	11833	8655	57.182
360	2490.8	1775.9	46.195	1700	12656	9280	57.680
380	2630.0	1875.3	46.571	1800	13486	9911	58.155
400	2769.1	1974.8	46.927	1900	14322	10549	58.607
420	2908.3	2074.3	47.267	2000	15164	11192	59.039
440	3047.5	2173.8	47.591	2100	16011	11841	59.451
460	3186.9	2273.4	47.900	2200	16863	12494	59.848
480	3326.5	2373.3	48.198	2300	17719	13151	60.228
500	3466.2	2473.2	48.483	2400	18579	13813	60.594
520	3606.1	2573.4	48.757	2500	19443	14479	60.946
537	3725.1	2658.7	48.982	2600	20311	15148	61.287
540	3746.2	2673.8	49.021	2700	21183	15821	61.616
560	3886.6	2774.5	49.276	2800	22058	16497	61.934
580	4027.3	2875.5	49.522	2900	22936	17177	62.242
600	4168.3	2976.8	49.762	3000	23818	17860	62.540
620	4309.7	3078.4	49.993	3100	24703	18546	62.831
640	4451.4	3180.4	50.218	3200	25591	19236	63.113
660	4593.5	3282.9	50.437	3300	26482	19928	63.386
680	4736.2	3385.8	50.650	3400	27376	20624	63.654
700	4879.3	3489.2	50.858	3500	28273	21323	63.914
720	5022.9	3593.1	51.059	3600	29174	22025	64.168
740	5167.0	3697.4	51.257	3700	30078	22730	64.415
760	5311.4	3802.2	51.450	3800	30984	23438	64.657
780	5456.4	3907.5	51.638	3900	31894	24149	64.893
800	5602.0	4013.3	51.821	4000	32806	24863	65.123
820	5748.1	4119.7	52.002	4100	33722	25580	65.350
840	5894.8	4226.6	52.179	4200	34640	26299	65.571
860	6041.9	4334.1	52.352	4300	35561	27022	65.788
880	6189.6	4442.0	52.522	4400	36485	27747	66.000
900	6337.9	4550.6	52.688	4500	37412	28475	66.208
920	6486.7	4659.7	52.852	4600	38341	29206	66.413
940	6636.1	4769.4	53.012	4700	39274	29940	66.613
960	6786.0	4879.5	53.170	4800	40209	30676	66.809
980	6936.4	4990.3	53.326	4900	41146	31415	67.003
1000	7088	5102	53.477	5000	42086	32157	67.193
1100	7850	5666	54.204	5100	43029	32901	67.380
1200	8626	6243	54.879	5200	43974	33648	67.562
1300	9413	6831	55.508	5300	44922	34397	67.743

Table B-6D

Ideal Gas Enthalpy, Internal Energy, and Absolute Entropy of Carbon Monoxide, CO
$\bar{h}$ and $\bar{u}$, Btu/lbmol; $\bar{s}'$, Btu/lbmol·°R

T, °R	$\bar{h}$	$\bar{u}$	$\bar{s}'$	T, °R	$\bar{h}$	$\bar{u}$	$\bar{s}'$
300	2081.9	1486.1	43.223	1400	9948	7168	54.129
320	2220.9	1585.4	43.672	1500	10711	7732	54.655
340	2359.9	1684.7	44.093	1600	11483	8306	55.154
360	2498.8	1783.9	44.490	1700	12264	8888	55.628
380	2637.9	1883.3	44.866	1800	13053	9479	56.078
400	2776.9	1982.6	45.223	1900	13850	10077	56.509
420	2916.0	2081.9	45.563	2000	14653	10682	56.922
440	3055.0	2181.2	45.886	2100	15463	11293	57.317
460	3194.0	2280.5	46.194	2200	16279	11911	57.696
480	3333.0	2379.8	46.491	2300	17101	12534	58.062
500	3472.1	2479.2	46.775	2400	17927	13161	58.414
520	3611.2	2578.6	47.048	2500	18759	13794	58.754
537	3729.5	2663.1	47.272	2600	19594	14431	59.081
540	3750.3	2677.9	47.310	2700	20434	15072	59.398
560	3889.5	2777.4	47.563	2800	21277	15717	59.705
580	4028.7	2876.9	47.807	2900	22124	16365	60.002
600	4168.0	2976.5	48.044	3000	22973	17016	60.290
620	4307.4	3076.2	48.272	3100	23826	17670	60.569
640	4446.9	3175.9	48.494	3200	24681	18326	60.841
660	4586.5	3275.8	48.709	3300	25539	18986	61.105
680	4726.2	3375.8	48.917	3400	26399	19647	61.362
700	4866.0	3475.9	49.120	3500	27262	20311	61.612
720	5006.1	3576.3	49.317	3600	28127	20978	61.855
740	5146.4	3676.9	49.509	3700	28994	21646	62.093
760	5286.8	3777.5	49.697	3800	29862	22316	62.325
780	5427.4	3878.4	49.880	3900	30733	22988	62.511
800	5568.2	3979.5	50.058	4000	31605	23662	62.772
820	5709.4	4081.0	50.232	4100	32479	24337	62.988
840	5850.7	4182.6	50.402	4200	33354	25014	63.198
860	5992.3	4284.5	50.569	4300	34231	25692	63.405
880	6134.2	4386.6	50.732	4400	35109	26371	63.607
900	6276.4	4489.1	50.892	4500	35989	27052	63.805
920	6419.0	4592.0	51.048	4600	36869	27734	63.998
940	6561.7	4695.0	51.202	4700	37751	28417	64.188
960	6704.9	4798.5	51.353	4800	38634	29102	64.374
980	6848.4	4902.3	51.501	4900	39518	29787	64.556
1000	6992	5006	51.646	5000	40403	30473	64.735
1100	7717	5532	52.337	5100	41289	31161	64.910
1200	8451	6068	52.976	5200	42176	31849	65.082
1300	9195	6613	53.571	5300	43063	32538	65.252

Table B-6E

Ideal Gas Enthalpy, Internal Energy, and Absolute Entropy of Carbon Dioxide, CO_2
$\bar{h}$ and $\bar{u}$, Btu/lbmol; $\bar{s}'$, Btu/lbmol·°R

T, °R	$\bar{h}$	$\bar{u}$	$\bar{s}'$	T, °R	$\bar{h}$	$\bar{u}$	$\bar{s}'$
300	2108.2	1512.4	46.353	1400	13345	10565	61.124
320	2256.6	1621.1	46.832	1500	14576	11597	61.974
340	2407.3	1732.1	47.289	1600	15829	12652	62.783
360	2560.5	1845.6	47.728	1700	17101	13725	63.555
380	2716.4	1961.8	48.148	1800	18392	14817	64.292
400	2874.7	2080.4	48.555	1900	19698	15925	64.999
420	3035.7	2201.7	48.947	2000	21019	17047	65.676
440	3199.4	2325.6	49.329	2100	22353	18182	66.327
460	3365.7	2452.2	49.698	2200	23699	19330	66.953
480	3534.7	2581.5	50.058	2300	25056	20489	67.557
500	3706.2	2713.3	50.408	2400	26424	21658	68.139
520	3880.3	2847.7	50.750	2500	27801	22837	68.702
537	4030.2	2963.8	51.032	2600	29187	24024	69.245
540	4056.8	2984.4	51.082	2700	30581	25219	69.771
560	4235.8	3123.7	51.408	2800	31983	26422	70.282
580	4417.2	3265.4	51.726	2900	33392	27633	70.776
600	4600.9	3409.4	52.038	3000	34807	28849	71.255
620	4786.8	3555.6	52.343	3100	36228	30072	71.722
640	4974.9	3704.0	52.641	3200	37655	31300	72.175
660	5165.2	3854.6	52.934	3300	39087	32533	72.616
680	5357.6	4007.2	53.225	3400	40524	33772	73.045
700	5552.0	4161.9	53.503	3500	41965	35015	73.462
720	5748.4	4318.6	53.780	3600	43411	36262	73.870
740	5946.8	4477.3	54.051	3700	44861	37513	74.267
760	6147.0	4637.9	54.319	3800	46314	38768	74.655
780	6349.1	4800.1	54.582	3900	47771	40026	75.033
800	6552.9	4964.2	54.839	4000	49231	41288	75.404
820	6758.3	5129.9	55.093	4100	50695	42553	75.765
840	6965.7	5297.6	55.343	4200	52162	43821	76.119
860	7174.7	5466.9	55.589	4300	53632	45093	76.464
880	7385.3	5637.7	55.831	4400	55015	46367	76.803
900	7597.6	5810.3	56.070	4500	56581	47645	77.135
920	7811.4	5984.4	56.305	4600	58060	48925	77.460
940	8026.8	6160.1	56.536	4700	59541	50208	77.779
960	8243.8	6337.4	56.765	4800	61025	51493	78.091
980	8462.2	6516.1	56.990	4900	62511	52781	78.398
1000	8682	6696	57.212	5000	64000	54071	78.698
1100	9803	7618	58.281	5100	65491	55363	78.994
1200	10955	8572	59.283	5200	66984	56658	79.284
1300	12137	9555	60.229	5300	68479	57954	79.569

Table B-6F

Ideal Gas Enthalpy, Internal Energy, and Absolute Entropy of Water, H_2O
$\bar{h}$ and $\bar{u}$, Btu/lbmol; $\bar{s}'$, Btu/lbmol°R

T, °R	$\bar{h}$	$\bar{u}$	$\bar{s}'$	T, °R	$\bar{h}$	$\bar{u}$	$\bar{s}'$
300	2367.6	1771.8	40.439	1400	11625	8845	53.168
320	2526.8	1891.3	40.952	1500	12551	9573	53.808
340	2686.0	2010.8	41.435	1600	13495	10318	54.418
360	2845.1	2130.2	41.889	1700	14455	11079	54.999
380	3004.4	2249.8	42.320	1800	15433	11858	55.559
400	3163.8	2369.4	42.728	1900	16428	12654	56.097
420	3323.2	2489.1	43.117	2000	17439	13467	56.617
440	3482.7	2608.9	43.487	2100	18467	14297	57.119
460	3642.3	2728.8	43.841	2200	19511	15142	57.605
480	3802.0	2848.8	44.182	2300	20571	16003	58.077
500	3962.0	2969.1	44.508	2400	21646	16880	58.535
520	4122.0	3089.4	44.821	2500	22735	17771	58.980
537	4258.3	3191.9	45.079	2600	23840	18676	59.414
540	4282.4	3210.0	45.124	2700	24957	19595	59.837
560	4442.8	3330.7	45.415	2800	26088	20528	60.248
580	4603.7	3451.9	45.696	2900	27231	21472	60.650
600	4764.7	3573.2	45.970	3000	28386	22429	61.043
620	4926.1	3694.9	46.235	3100	29553	23397	61.426
640	5087.8	3816.8	46.492	3200	30730	34375	61.801
660	5250.0	3939.3	46.741	3300	31918	25365	62.167
680	5412.5	4062.1	46.984	3400	33116	26364	62.526
700	5575.4	4185.3	47.219	3500	34324	27373	62.876
720	5738.8	4309.0	47.450	3600	35540	28391	63.221
740	5902.6	4433.1	47.673	3700	36765	29418	63.557
760	6066.9	4557.6	47.893	3800	37999	30453	63.887
780	6231.7	4682.7	48.106	3900	39240	31495	64.210
800	6396.9	4808.2	48.316	4000	40489	32546	64.528
820	6562.6	4934.2	48.520	4100	41745	33603	64.839
840	6728.9	5060.8	48.721	4200	43008	34668	65.144
860	6895.6	5187.8	48.916	4300	44278	35739	65.444
880	7062.9	5315.3	49.109	4400	45554	36816	65.738
900	7230.9	5443.6	49.298	4500	46836	37900	66.028
920	7399.4	5572.4	49.483	4600	48124	38989	66.312
940	7568.4	5701.7	49.665	4700	49417	40083	66.591
960	7738.0	5831.6	49.843	4800	50716	41183	66.866
980	7908.2	5962.0	50.019	4900	52019	42288	67.135
1000	8079	6093	50.191	5000	53327	43398	67.401
1100	8942	6758	51.013	5100	54640	44512	67.662
1200	9820	7437	51.777	5200	55957	45631	67.918
1300	10715	8133	52.494	5300	57279	46754	68.172

Table B-6G

Ideal Gas Enthalpy, Internal Energy, and Absolute Entropy of Diatomic Hydrogen, H_2
$\bar{h}$ and $\bar{u}$, Btu/lbmol; $\bar{s}'$, Btu/lbmol°R

T, °R	$\bar{h}$	$\bar{u}$	$\bar{s}'$	T, °R	$\bar{h}$	$\bar{u}$	$\bar{s}'$
300	2063.5	1467.7	27.337	1400	9674	6894	37.883
320	2189.4	1553.9	27.742	1500	10382	7403	38.372
340	2317.2	1642.0	28.130	1600	11093	7915	38.830
360	2446.8	1731.9	28.501	1700	11807	8431	39.264
380	2577.8	1823.2	28.856	1800	12527	8952	39.675
400	2710.2	1915.8	29.195	1900	13251	9478	40.067
420	2843.7	2009.6	29.520	2000	13980	10008	40.441
440	2978.1	2104.3	29.833	2100	14715	10544	40.799
460	3113.5	2200.0	30.133	2200	15454	11086	41.143
480	3249.4	2296.2	30.424	2300	16200	11632	41.475
500	3386.1	2393.2	30.703	2400	16951	12185	41.794
520	3523.3	2490.6	30.972	2500	17707	12743	42.104
537	3640.3	2573.9	31.194	2600	18470	13306	42.403
540	3660.9	2588.5	31.232	2700	19238	13876	42.692
560	3798.8	2686.7	31.482	2800	20012	14451	42.973
580	3937.1	2785.3	31.724	2900	20792	15033	43.247
600	4075.6	2884.1	31.959	3000	21577	15619	43.514
620	4214.3	2983.1	32.187	3100	22368	16212	43.773
640	4353.1	3082.1	32.407	3200	23164	16809	44.026
660	4492.1	3181.4	32.621	3300	23966	17412	44.273
680	4631.1	3280.7	32.829	3400	24772	18020	44.513
700	4770.2	3380.1	33.031	3500	25583	18632	44.748
720	4909.5	3479.6	33.226	3600	26399	19249	44.978
740	5048.8	3579.2	33.417	3700	27219	19871	45.203
760	5188.1	3678.8	33.603	3800	28043	20497	45.423
780	5327.6	3778.6	33.784	3900	28871	21126	45.638
800	5467.1	3878.4	33.961	4000	29704	21760	45.849
820	5606.7	3978.3	34.134	4100	30540	22398	46.056
840	5746.3	4078.2	34.302	4200	31380	23039	46.257
860	5885.9	4178.0	34.466	4300	32224	23684	46.456
880	6025.6	4278.0	34.627	4400	33071	24333	46.651
900	6165.3	4378.0	34.784	4500	33922	24985	46.842
920	6305.1	4478.1	34.938	4600	34776	25641	47.030
940	6444.9	4578.1	35.087	4700	35633	26299	47.215
960	6584.7	4678.3	35.235	4800	36493	26961	47.396
980	6724.6	4778.4	35.379	4900	37357	27626	47.574
1000	6865	4879	35.520	5000	38223	28294	47.749
1100	7565	5380	36.188	5100	39093	28965	47.921
1200	8266	5883	36.798	5200	39965	29639	48.090
1300	8969	6387	37.360	5300	40840	30315	48.257

Table B-7

Values of the Enthalpy of Formation, $\bar{h}_F$, Gibbs Function of Formation, $\bar{g}_F$, and Absolute Entropy, at 25°C (77°F) and 1 atm[a]

Substance	Formula	$\bar{h}_F$ Btu/lbmol	$\bar{g}_F$ Btu/lbmol	Absolute Entropy Btu/(lbmol)(°R)
Carbon	C(s)	0	0	1.36
Hydrogen	H_2(g)	0	0	31.21
Nitrogen	N_2(g)	0	0	45.77
Oxygen	O_2(g)	0	0	49.00
Carbon monoxide	CO(g)	− 47,550	− 59,010	47.21
Carbon dioxide	CO_2(g)	−169,300	−169,680	51.07
Water	H_2O(g)	−104,040	− 98,350	45.11
Water	H_2O(l)	−122,970	−102,040	16.71
Methane	CH_4(g)	− 32,210	− 21,860	44.49
Acetylene	C_2H_2(g)	+ 97,540	+ 89,990	48.00
Ethylene	C_2H_4(g)	+ 22,490	+ 29,310	52.45
Ethane	C_2H_6(g)	− 36,430	− 14,150	54.85
Propane	C_3H_8(g)	− 44,680	− 10,110	64.51
n-Octane	C_8H_{18}(g)	− 89,680	+ 7,110	111.55
n-Octane	C_8H_{18}(l)	−107,530	+ 2,840	86.23

[a]From the JANAF Thermochemical Tables, Dow Chemical Co., 1971; *Selected Values of Chemical Thermodynamic Properties*, NBS Tech Note 270–3, 1968; and *API Research Project 44*, Carnegie Press, 1953.

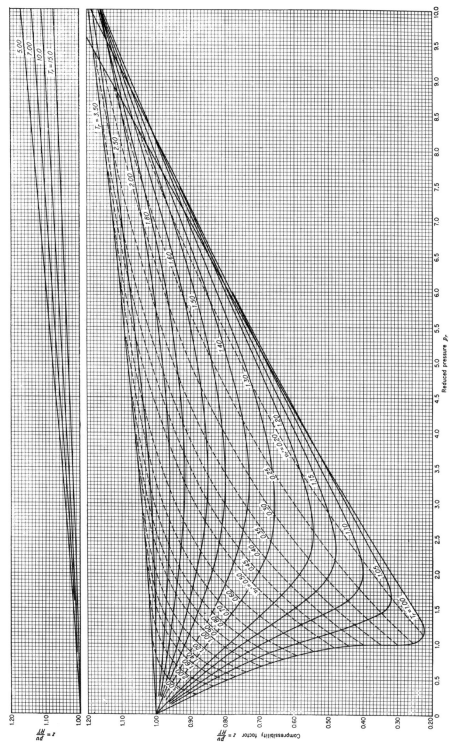

Figure B-1 Generalized compressibility chart, $T_r = T/T_c$, $p_r = p/p_c$, $v_r' = vp_c/RT_c$. (From L. C. Nelson and E. F. Obert, "Generalized p-V-T Properties of Gases," Trans. ASME, **76**, October 1954, 1057–1066.)

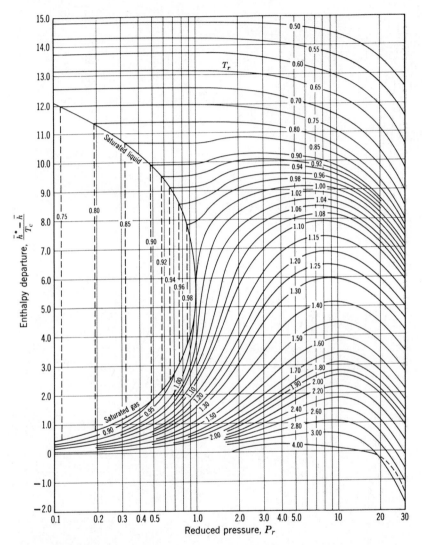

Figure B-2 Generalized enthalpy departure chart. (From G. J. Van Wylen and R. E. Sonntag, "Fundamentals of Classical Thermodynamics," 2nd ed., John Wiley & Sons, New York, 1973.)

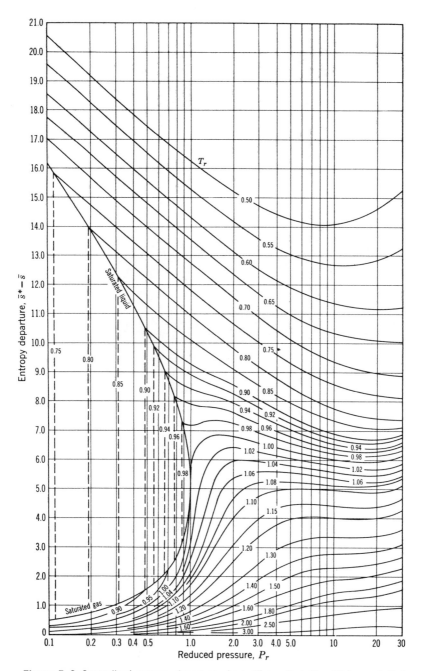

Figure B-3 Generalized entropy departure chart. (From G. J. Van Wylen and R. E. Sonntag, "Fundamentals of Classical Thermodyamics," 2nd ed., John Wiley & Sons, New York, 1973.)

Index

Author Index

Ahrendts, J., 173
Angus, S., 181

Babu, S.P., 176
Baker, C.R., 82
Bejan, A., 79, 81, 84
Bird, R.B., 43, 84
Blackshear, Jr., P.L., 100
Bosnjakovic, F., 100
Bruges, E.A., 76
Brzustowski, T.A., 104, 173

Canada, J.R., 216
Clarke, J.M., 198
Crellin, G.L., 216

DeGarmo, E.P., 216
de Reuck, K.M., 181

Edgerton, R.H., 198
El-Sayad, Y.M., 143, 216
Evans, R.B., 143, 216(2)
Evans, W.H., 155

Fabrycky, W.J., 216
Faires, V.M., 39
Fazzolare, R.A., 173, 216
Fehring, T., 216

Gaggioli, R.A., 76, 143, 173(3), 198, 216(3)
Gatts, R.R., 7
Golem, P.J., 104, 173
Greenkorn, R., 225
Gyftopoulos, E.P., 8(2), 39, 87

Hall, E.H., 8, 87
Hatsopoulos, G.N., 8, 39
Hawkins, G.A., 39
Haywood, R.W., 76, 143

Hildebrand, F.B., 78
Hill, P.G., 226
Horlock, J.H., 198
Hougen, O., 225

Jaffe, I., 155
Jones, J.B., 39

Kaye, J., 243
Keenan, J.H., 39(2), 76(2), 120, 226, 243
Keyes, F.G., 226
Kobe, K.A., 242
Kotas, T.J., 100
Kreider, J.F., 216
Krieth, F., 68(2), 69, 81
Kuehn, T.H., 198

Landsberg, P.T., 198
Lazaridis, L.J., 8, 87
Levine, S., 155
Lightfoot, E.N., 43, 84
Lydersen, A., 225

Massey, R.G., 7
Moore, J.G., 226
Moran, M.J., 121

Nelson, L.C., 251

Obert, E.F., 39, 76, 143, 216, 242, 251

Parrott, J.E., 198
Perkins, H.C., 39
Petela, R., 198
Petit, P.J., 173
Press, W.H., 198

Rant, Z., 164
Reid, R.C., 120, 192
Reistad, G.M., 83, 101, 120(2)
Reynolds, W.C., 39, 195

Riekert, L., 173
Robertson, J.C., 7
Rodriquez, L., 198
Rossini, F.D., 155

Sant, R.W., 8
Shaner, R.L., 82
Shapiro, A.H., 39
Shapiro, H.N., 198
Sherwood, T.K., 120, 192
Shinskey, F.G., 8
Simmang, C.M., 39
Singh, S.P., 176
Smith, C.B., 8, 173, 216
Sonntag, R.E., 39, 252, 253
Spanner, D.C., 198
Spiegler, K.W., 216
Stewart, W.E., 43, 84
Strobridge, T.R., 240
Styrylska, T., 198

Su, G., 33
Sullivan, W.G., 216
Sussman, M.V., 173
Szargut, J.; 173, 198

Tapia, C.F., 121
Thirumaleshwar, M., 121
Thuesen, G.J., 216
Thuesen, H.G., 216
Trepp, C., 198
Tribus, M., 216

Van Wylen, G.J., 39, 252, 253

Wagman, D.D., 155
Wark, K., 39
Weil, S.A., 176
Wepfer, W.J., 143, 173, 216
Widmer, T.F., 8(2), 87
Wolfe, H.C., 101

Subject Index

Absolute entropy, 150
  tables, 244–50
Additive partial pressures, Dalton's law, 37
Adiabatic process, 12
Air:
  air-water vapor mixtures, 139
  molecular weight, 139
  table of ideal gas properties, 243
  theoretical amount of, 147
Air preheating, in combustion, 165
Annualized costs, 204-6 (see also Cost equations)
Asset, 201
Asset life, 201
Availability, 2, 7, 122, 129, 131, 146, 171
  chemical (see Chemical availability)
  destruction (see Availability destruction)
  equation:
    control mass, 60
    control volume, 65
  flow (see Flow availability)
  flow with heat, 58, 60
  flow with work, 59, 60
  loss, general comment, 7, 66, 86
  marginal cost, 214
  thermomechanical (see Thermomechanical availability)
  unit cost, 206
  use in evaluating chemical processes, 169–71
  use in radiation heat transfer, 194–97
  use in thermoeconomics (see Thermoeconomics)
Availability destruction, 2, 7, 55, 86 (see also Entropy production; Irreversibility)
  control mass, 60
  control volume, 65
Availability equation:
  control mass, 60
  control volume, 65
  power cycle, 93
  reversed cycle, 96

$b (= h - T_0 s)$, 70, 107
  property figure for water, 108
Back pressure turbine, 6, 174
Boundary, system, 9

Calculus of variations, 78
Capital-recovery factor, 203-6
  table, 204
Carbon, chemical availability, 155, 156
Carbon dioxide, table of ideal gas properties, 247

Carbon monoxide:
  chemical availability, 158
  table of ideal gas properties, 246
Carbon-steam process, 166, 176
Carnot efficiency, 22
Celsius temperature, 220
Chemical availability, 122, 129, 132, 146, 153, 169, 172
  carbon monoxide, 158
  fuel (see Fuel chemical availability)
  standard chemical availability, 170
    reference substances, 170
    table for selected elements, 171
Chemical equilibrium, 11, 126, 153
Chemical potential, 124
  in mixtures, 127-28
  of pure components, 125
Chemical processes, availability analysis, 169–71
Clausius inequality, 22
Clausius postulate, 18
Coal gasification, 166–68, 175, 176
Coefficient of performance:
  heat pump cycle, 96
  refrigeration cycle, 97
Combined system, 46, 130, 171
Combustion, 147
  complete, 147
  irreversibility, 159
Compressed liquid, 31
  table for water, 234
Compressibility chart, generalized, 32, 33, 251
Compressibility factor, 32
  table of critical values, 225
Condenser cooling water, availability, 162-63
Conduction heat transfer, 43, 77
  thermal conductivity, 43, 68, 77, 84
Conservation of mass, 15
Continuity equation, 42
Control mass, 10
  availability destruction, 60
  availability equation, 60
  energy equation, 14
  entropy equation, 26
  entropy production, 25
Control surface, 9
  adiabatic, 14
Control volume, 10
  availability destruction, 65
  availability equation, 65
  energy equation, 18
  entropy equation, 28
  entropy production, 27

Index

Convection heat transfer, 68
  coefficient, 68
  forced, 68
  free, 68
Conversion factors, table, 220
Cooling tower, 141
Cost, 200
  annualized, 204-6
  classifications (see Cost classifications)
  equations (see Cost equations)
Cost classifications:
  first, 200
  fixed, 200
  incremental, 200
  marginal, 200
    of availability, 214
  sunk, 201
  total, 200
  unit, 200
    of availability, 206
  variable, 200
Cost equations, 205 (see also Annualized costs)
  local, 215
  use in costing, 206-10
  use in design, 210-15
Critical point, 31
  data, 225
Critical pressure, 31
  data, 225
Critical properties, table, 225
Critical temperature, 31
  data, 225
Cryogenic refrigerator, 180
Cycle (see also Reversed cycles):
  heat pump, 82, 96, 102
  power (see Heat engine)
  refrigeration, 97
  reversed, 95

Dalton mixture model, 36
Dead state, 122, 137
  restricted (see Restricted dead state)
Departure:
  enthalpy, 116
    generalized chart, 252
  entropy, 117
    generalized chart, 253
  flow availability, 119
Design variables, 211-15
  local, 214
Differential, total or exact, 39, 124
  test for exactness, 39
Dry bulb temperature, 141

Earning power of money, 202
Efficiency, 85-86
  Carnot, 22
  isentropic turbine, 74
  second law (see Second law efficiency)
  thermal, 15
Endothermic reaction, 166, 176
Energy, 2, 13
  equation (see Energy equation)
  internal (see Internal energy)
  kinetic, 13
  mechanical, 42
  potential, 13
  quantity versus potential (quality), 3, 7, 44
  utilization, 3-7
    barriers to improving efficiency of use, 3-4
    bottoming cycle, 6
    cogeneration, 6
    energy management programs, 5
    housekeeping, 5
    leak-plugging, 5
    multiple use methods, 5

Energy (cont.)
  power recovery, 5
  primary end uses in industry, 5
  topping cycle, 6
  waste heat recovery, 5
Energy equation:
  control mass, 14
  control volume, 18
  cycle, 14
Enthalpy, 13
  of combustion, 150
  departure, 116
    generalized chart, 252
  of formation, 148
    table, 250
  ideal gas, 35
    tables, 36, 243-49
  ideal gas mixtures, 37
  incompressible liquids, 38
  partial molal, 128
Entropy, 23
  absolute, 150
    tables, 244-50
  departure, 117
    generalized chart, 253
  equation (see Entropy equation)
  flow with heat, 26
  flow with work, 26
  ideal gas, 35
    tables, 36, 243-49
  ideal gas mixtures, 37
  incompressible liquids, 38
  increase of entropy principle, 24
  partial molal, 128
  production (see Entropy production)
  radiation, 195
Entropy equation:
  control mass, 26
  control volume, 28
Entropy production (see also Availability destruction; Irreversibility):
  control mass, 25
  control volume, 27
Environment, 44, 129, 135, 140, 146, 154, 169, 170, 173, 178, 179, 182, 186, 191 (see also Dead state; Restricted dead state)
Equation of state, 30
Equilibrium, 11
  chemical, 11, 126, 153
  criteria, 25, 125
  mechanical, 11, 126
  membrane, 129
  phase, 125, 136
  state, 11
    mutual stable, 12
    stable, 12
  thermal, 11, 126
Essergy, 131
Exact differential (see Differential, total or exact)
Exergy, 7, 64, 131
Exothermic reaction, 147
Extensive property, 10

Fahrenheit temperature, 220
First law of thermodynamics, 12
Flow availability, 134, 172
  chemical availability, 134 (see also Chemical availability)
  contours on a psychrometric chart, 141
  of ideal gas mixtures, 137
  of moist air, 140
  of products of combustion, 138, 160, 163, 167, 187, 192
  thermomechanical flow availability, 64, 105-21
    departure, 119
    from generalized property charts, 116-20

Flow availability (*cont.*)
  ideal gas, 111–13
  ideal gas mixtures, 113–15
  property figures:
    nitrogen, 110
    Refrigerant 12, 109
    water, 110
  reduced, 113
    property figure, ideal gas with constant specific heat, 114
  of water streams, 135
Flow work, 17
Fourier's heat conduction model, 43, 77
Friction factor, 69, 213
Fuel cell, 152, 162
Fuel chemical availability, 146–79
  approximations, 156, 174, 192
  evaluation, 153, 158, 169
  of hydrocarbons, 154, 156, 173, 186
    table, 155
  standard, 170
    reference substances, 170
    table for selected elements, 171
Furnace, gas-fired, 156, 177

g_c, 13, 221
Gas constant (*see* Universal gas constant)
Gas liquefaction, 82, 180
Gas scale of temperature, 11
Gas turbine power plant, 66, 101
Generalized property charts:
  compressibility, 32, 33, 251
  enthalpy, 116, 252
  entropy, 117, 253
Gibbs free energy (*see* Gibbs function)
Gibbs function, 29 (*see also* Chemical potential)
  of formation, 151
    table, 250
  ideal gas mixtures, 127
  partial molal, 128
Grashof number, 68

Heat, 13
  capacity (*see* Specific heat)
  conduction (*see* Conduction heat transfer)
  convection (*see* Convection heat transfer)
  flow of availability with, 58, 60
  flow of entropy with, 26
  flux, 28
  interactions, 12
  radiation (*see* Radiation)
  sign convention, 14
Heat engine, 14
  availability equation, 93
  energy equation, 14
  reversible, 21, 52
  thermal efficiency, 15
Heat exchanger effectiveness, 81
Heating values, 150
  table for selected hydrocarbons, 155
Heat pump, 82, 96, 102 (*see also* Reversed cycles)
Heat transfer coefficient, convection, 68
Humidification, 145
Humidity ratio, 139
Hydrocarbon fuels (*see also* Fuel chemical availability):
  chemical availability, 146, 154, 156, 173, 186
    approximations, 156, 174, 192
    table, 155
  combustion, 147
  heating values, table, 155
Hydrogen:
  chemical availability, 155, 173
  table of ideal gas properties, 249

Ideal gas mixtures, 36
  availability, 113, 132, 137 (*see also* Flow availability; Thermomechanical availability)
  chemical potential, 127
  enthalpy, 37
  entropy, 37
  Gibbs function, 127
  internal energy, 37
  molecular weight, 37
  specific heat, 37
Ideal gas model, 34
  availability (*see* Flow availability; Thermomechanical availability)
  enthalpy (*see* Enthalpy)
  entropy (*see* Entropy)
  internal energy (*see* Internal energy)
  mixtures (*see* Ideal gas mixtures)
  specific heat relations, 35
    table, 242
  tables for selected gases, 243–49
Incompressible liquids, 37
  property relations, 38
Inflation, 202 (*see also* Time value of money)
Intensive property, 10
Interest, 202 (*see also* Time value of money)
  compound, 202
  effective rate, 216
  rate, 202
Internal energy, 13
  ideal gas, 35
    tables for selected gases, 36, 243–49
  ideal gas mixtures, 37
  incompressible liquids, 38
  radiation, 195
Internally reversible process, 20
International System of units (SI), 219
Intrinsic property, 10
Irreversibilities, 20
  external, 20
  internal, 20
Irreversibility, 55, 69 (*see also* Availability destruction; Entropy production)
  combustion irreversibility, 159
Irreversible process, 19
Isentropic process, 49
Isentropic turbine efficiency, 74

Kay's mixture rule, 120
Kelvin-Planck statement of the second law, 18
Kelvin temperature scale, 11, 21
Kinetic energy, 13
Kinetic theory, 9, 195

Lagrange multipliers, 214, 222
Laminar flow, 84
Life-cycle costing, 202
Liquefaction of gases, 82, 180

Marginal cost, 200
  of availability, 214
Mass flow rate, 15
Maxwell relation, 40
Mechanical equilibrium, 11, 126
Methane:
  chemical availability, 155, 186
  combustion, 147
  heating values, 155
Method of Undetermined Lagrange Multipliers, 214, 221
Mixtures:
  Dalton mixture model, 36
  ideal gas, 36–37, 127
  mixture rules, 120
  partial molal properties, 128
Mole, 10
Mole fraction, 36

Index

Molecular weight, 10
  air, 139
  ideal gas mixtures, 37
  water, 139
Multicomponent systems, 123–28

Newton's second law of motion, 221
Nitrogen:
  flow availability diagram, 110
  National Bureau of Standards tables, 240–241
  table of ideal gas properties, 244
Nusselt number, 68, 213

One-dimensional flow, 15
Optimal design, 210
Overall unit conductance, 104
Oxygen, table of ideal gas properties, 245

Partial molal properties, 128
Partial pressure, 36
Path, 20
Perpetual-motion machine of the second kind (PMM2), 18
Phase:
  equilibrium, 125, 136
  liquid, 31
  solid, 31
  two-phase regions, 31
  vapor (gas), 31
Photons, 195
Photosynthesis, 196
Potential energy, 13
Power cycle (*see* Heat engine)
Prandtl number, 68
Present worth, 203
Present-worth factor, 203
Pressure:
  critical, 31
    table, 225
  partial, 36
  pseudocritical, 120
  radiation, 195
  reduced, 32
Price, 200
Process, 11
  adiabatic, 12
  internally reversible, 20
  irreversible, 19
  isentropic, 49
  reversible, 19
  throttling, 71, 98, 136
Products of combustion, 147
  flow availability, 138, 160, 163, 167, 187, 192
Property, 10
  critical, 31
    table, 225
  extensive, 10
  intensive, 10
  intrinsic, 10
  partial molal, 128
  pseudocritical, 120
  reduced, 32
  specific, 10
  thermostatic, 10
Pseudocritical pressure, 120
Pseudocritical temperature, 120
Psychrometric chart, 107, 141
Psychrometric processes, 139–43
Purchasing power of money, 202

Quality, 31

Radiation, 68, 194 (*see also* Sunlight)
  isotropic, 195
  solar, 196

Rankine temperature scale, 11
Reactants, 147
Reduced pressure, 32
Reduced temperature, 32
Reference substances, for standard chemical availabilities, 170, 171
Refrigerant 12:
  Du Pont Freon Products Division table, 235–39
  flow availability diagram, 109
Refrigeration:
  cryogenic, 180
  cycle, 97 (*see also* Reversed cycles)
  Helium system, 181
Refuse, 160, 193
Relative humidity, 139
Restricted dead state, 45, 56, 171
  influence in availability diagrams, 107
Reversed cycles, 95
  availability equation, 96
  coefficients of performance, 96, 97
  energy equation, 96
Reversible process, 19
  internally, 20
Reynolds number, 68, 213
Roughness factor, 69, 213

Salvage value, 201
Saturated liquid, 31
Saturated vapor, 31
Saturation line, 31
Saturation pressure, 31
Saturation state, 31
Saturation temperature, 31
Second law efficiency, 85–104
  exergetic efficiency, 85
  local, 104
  rational efficiency, 85
  task, 86
  values for selected U.S. industries, 87
  thermodynamic efficiency ratio, 85
  turbine effectiveness, 74, 98
  uses of, 87
Second law of thermodynamics, 18
  Clausius postulate, 18
  corollaries, 21
  Kelvin-Planck statement, 18
Semipermeable membranes, 128
SI (International System) units, 219
Sign convention:
  heat, 14
  work, 12
Simple compressible substance, 28
  property relations, 28, 123, 147
  state principle, 28
Solid waste recovery system, 188
Specific heat, 29
  ideal gas, 35, 112
    table, 242
  ideal gas mixtures, 37
  incompressible liquids, 38
Specific heat ratio, 29
Specific property, 10
Standard chemical availabilities, 170
  reference substances, 170, 171
  table, 171
State, 10
  datum, in property tables, 32, 148
  dead (*see* Dead state)
  equilibrium, 11
    mutual stable, 12
    stable, 12
  principle, for simple compressible substances, 28
  restricted dead state (*see* Restricted dead state)
  saturation, 31
  steady-state, 15
State principle, for simple compressible substances, 28

Steady-state, 15
Steam generator, 157, 163, 189
Stefan-Boltzman radiation law, 195
Stoichiometric coefficients, 147
Substitute natural gas (SNG), 175
Sunlight:
    availability of, 196–97
    diffuse, 196
    direct, 196
Superheated vapor, 31
    table for nitrogen, 241
    table for Refrigerant 12, 237–39
    table for water, 230–33
Surroundings, 10
System, 9
    closed, 10
    combined, 46, 130, 171
    isolated, 18, 125
    multicomponent, 123
    open, 10
    subsystem, 213

TdS equations, 29, 123, 124
Temperature, 10
    Celsius scale, 220
    critical, 31
        table, 225
    dry bulb, 141
    empirical scales, 11
    Fahrenheit scale, 220
    gas scale, 11
    Kelvin scale, 11, 21
    pseudocritical, 120
    Rankine scale, 11
    reduced, 32
    wet bulb, 141
Theoretical amount of air, 147
Thermal conductivity, 43, 68, 77, 84
Thermal efficiency, heat engine, 15
Thermal energy reservoir (TER), 20
Thermal equilibrium, 11, 126
Thermal radiation (see Radiation)
Thermodynamics, 9
    classical, 9
    first law, 12
    second law, 18
    third law, 150
Thermoeconomics, 199–218
    annualized costs, 204–6
    costing, 206–10
    design, 210–15
    subsystem analysis, 213–15
    use of second law efficiencies, 207, 209, 210, 215, 217
Thermomechanical availability, 44–84, 122, 129, 132, 146, 171
    flow availability, 64, 105–21
        departure, 119
        figure for nitrogen, 110

Thermomechanical availability (cont.)
        figure for Refrigerant 12, 109
        figure for water, 110
        from generalized property charts, 116–20
        ideal gas, 111–13
        ideal gas mixtures, 113–15
    from property tables, 105–7
    property diagrams, 107
    property relations, 105–21
    reduced availability and reduced flow availability, 113
        figures, ideal gas with constant specific heat, 114–15
Thermometric property, 10
Thermostatic property, 10
Third law of thermodynamics, 150
Throttling process, 71, 98, 136
Time value of money, 202 (see also Interest)
    earning power, 202
    inflation, 202
    purchasing power, 202
Total differential (see Differential, total or exact)
Triple line, 30
Triple point, 30
Turbine effectiveness, 74, 98 (see also Second law efficiency)
Turbojet engine, 184

Unit cost, 200
    of availability, 206
United States Customary System of Units (USCS), 219
Units:
    International System (SI), 219
    selected conversions, 220
    United States Customary System (USCS), 219
Universal gas constant, 11, 34, 220
    specific gas constant, 34

van't Hoff equilibrium box, 152
Vapor power plant, 6, 160–66, 174, 176
Viscosity, 43, 84

Water:
    $b (= h - T_0 s)$ diagram, 108
    flow availability, 135
    Keenan, Keyes, Hill, and Moore "Steam Tables", 226–34
    molecular weight, 139
    table of ideal gas properties, 248
    thermomechanical flow availability diagram, 110
Wet bulb temperature, 141
Work, 12
    flow of availability with, 59, 60
    flow of entropy with, 26
    flow work, 17
    interactions, 12
    sign convention, 12

Date Due